# The Society of Cells

## *Cancer and Control of Cell Proliferation*

To our parents, sisters and Javier

# The Society of Cells

## *Cancer and Control of Cell Proliferation*

**C. Sonnenschein and A.M. Soto**

*Tufts University School of Medicine, Department of Anatomy and Cellular Biology, Boston, MA 02111, USA*

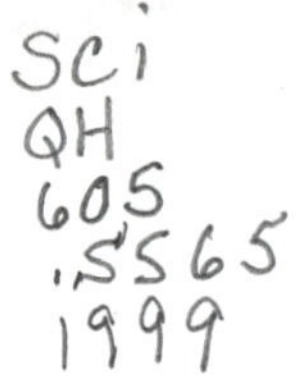


C. Sonnenschein and A.M. Soto
*Tufts University School of Medicine, Department of Anatomy and Cellular Biology, 136 Harrison Avenue, Boston, MA 02111, USA*

**Published in the United States of America, its dependent territories and Canada by arrangement with BIOS Scientific Publishers Ltd, 9 Newtec Place, Magdalen Road, Oxford OX4 1RE, UK**



First published 1999



A CIP catalogue record for this book is available from the British Library.

**Library of Congress Cataloging-in-Publication Data**
Sonnenschein, Carlos.
The society of cells: control of cell proliferation and cancer/Carlos Sonnenschein, Ana M. Soto.
p. cm.
Includes bibliographical references and index.
ISBN 0–387–91583–4 (pbk.: alk. paper)
1. Cell proliferation. 2. Cancer cells–Proliferation. 3. Carcinogenesis.
I. Soto, Ana M. II. Title.
QH605.S565 1998
616.99′4071–dc21 98–40928
CIP

ISBN 0-387-91583-4 Springer-Verlag New York Berlin Heidelberg
SPIN 19901509

Springer-Verlag New York Inc.
175 Fifth Avenue, New York
NY 10010-7858, USA

Production Editor: Andrea Bosher
Typeset by J&L Composition Ltd, Filey, UK
Printed by TJ International, Padstow, UK

# Contents

**Abbreviations** viii
**Preface** ix
**Acknowledgements** xiii

**PART 1**

1. **Cell proliferation: the background and the premises** **1**
Introduction 1
A brief history 2
Understanding the control of cell proliferation 7
The hierarchical organization of nature 9
Conclusions 11
References 12

2. **Cell proliferation, cell nutrition and evolution** **14**
Evolutionary perspective on the control of cell proliferation 14
The role of nutrition on the control of cell proliferation 20
Conclusions 26
References 27

3. **The experimental evaluation of cell proliferation** **31**
Introduction 31
Fitting data into working hypotheses 37
The use and misuse of tritiated thymidine incorporation data 38
Conclusions 39
References 40

4. **Hypotheses for the control of cell proliferation** 41
Historical perspective 41
Positive hypotheses 45
Negative hypotheses 52
Conclusions 55
References 55

5. **Sex hormone-mediated control of cell proliferation** 60
Introduction 60
Control of cell proliferation by estrogens 61
Hypotheses 62
Control of cell proliferation by androgens 71
Conclusions 72
References 74

6. **Cell proliferation and tissue differentiation** 78
The different meanings of 'differentiation' 78
A hierarchical perspective 79
The elusive concept of stem cells 80
The hemopoietic stem cells 81
Proliferation of antibody-producing cells: a different perspective 84
Conclusions 86
References 88

**PART 2**

7. **Introduction to carcinogenesis and neoplasia** 91
If normal cells beget normal cells, and neoplastic cells beget neoplastic cells, what causes normal cells to become neoplastic? 93
Germ-line mutations and carcinogenesis 95
Somatic mutations, cell proliferation, cell death, and the 'differentiation' hypotheses 96
Are carcinogenesis and neoplasia cellular or tissue-based phenomena? 97
References 98

| | | |
|---|---|---|
| **8.** | **The enormous complexity of cancer** | **99** |
| | Introduction | 99 |
| | Hierarchical levels of complexity in cancer | 101 |
| | Conclusions | 108 |
| | References | 109 |
| **9.** | **Facts and fantasies in carcinogenesis** | **112** |
| | Introduction | 112 |
| | Experimental carcinogenesis: the lessons missed | 118 |
| | Foreign-body carcinogenesis | 118 |
| | Physical carcinogenesis | 120 |
| | Chemical carcinogenesis | 120 |
| | Viral carcinogenesis. But . . . are viruses *per se* carcinogenic agents? | 122 |
| | On the uniqueness of the neoplastic cell | 125 |
| | Carcinogenesis in a flask | 127 |
| | Conclusions | 128 |
| | References | 130 |
| **Epilogue.** | **Moving toward the integration of cell proliferation, carcinogenesis, and neoplasia into biology** | **134** |
| | Ideology, silent assumptions, and operational definitions | 135 |
| | The universality of *proliferation* as the default state of all cells | 137 |
| | Forerunners of the tissue organization field hypothesis | 139 |
| | On paradigm changes | 139 |
| | An integrative approach of control of cell proliferation and carcinogenesis | 140 |
| | The impact of our reassessment | 142 |
| | And . . . finally | 142 |
| | References | 143 |
| | **Index** | **145** |

# Abbreviations

| | |
|---|---|
| BCSG | brain cell surface glycopeptide |
| CNS | central nervous system |
| CSF | colony-stimulating factor |
| EGF | epidermal growth factor |
| EPO | erythropoietin |
| FGF | fibroblast growth factor |
| FGFR | fibroblast growth factor receptor |
| HGF | hepatocyte growth factor |
| IL-2 | interleukin-2 |
| KGF | keratinocyte growth factor |
| MDGI | mammary-derived growth inhibitor |
| NGF | nerve growth factor |
| Rb | retinoblastoma |
| SCF | stem cell factor |
| SV40 | Simian Virus 40 |
| TGF | transforming growth factor |

# Preface

While you are browsing through this book, billions of cells in your body are actively proliferating. At the same time, probably an equal amount of cells, if not more, are not. Why? This book delves into this question and into the medical problem that is most closely related to it, cancer.

For most of our professional careers as physician-scientists, we have explored how sex hormones (estrogens and androgens) regulate the proliferation of their target cells in the reproductive tract of mammals, which is the site of the very organs that are prone to develop hormone-related cancers. Intellectually, we were born and bred in the midst of the molecular biology revolution and, as bench scientists, use this powerful methodology in search of the inhibitors of cell proliferation.

The lack of collective progress toward understanding the control of cell proliferation and cancer prompted us to embark on a parallel intellectual journey into the 'other' biologies: evolutionary, developmental, and organismal. We learned to ask 'why' questions, in addition to the 'how' and 'what' questions that normally concerned us and our peers in the pursuit of biochemical answers to biological questions. This parallel intellectual pursuit was started with trepidation from reading about the history and epistemology of biological thought, and from searching the literature pertaining to the theories on the control of cell proliferation and carcinogenesis. We realized that neither the history nor the epistemology of these important topics in biology had been researched fully. Hence, we explored the primary literature. Again, we found no explicit mention of the basic premises assumed by the scientists in doing the work that we were reviewing. At first, we were surprised by this finding; however, it appears to be a common occurrence. In the words of Mayr, '[Scientists] rarely articulate—if they think of it at all—what truths or concepts they accept without further question and what others they totally reject'. Thus, we felt two disparate emotions, one of inadequacy, as invaders of disciplines not our own, and one of elation, because our analysis, in addition to being novel, resulted in explanations that were not trivial, and in predictions that were different from those reached by the consensus of our peers. Most importantly, this reassessment generated new insights that reoriented our research program.

Thus, we felt that writing a book about what we learned through this personal exploratory journey would be useful to our students and colleagues alike. This is a book about ideas, and the data gathered to explore them. It has been in the making for over 10 years. During this period, we changed the scope of the book and the readers we wanted to reach several times. Each time, the content and the details in the book changed accordingly. Finally, we decided that the main purpose of the book was to address the problems, that in our view, have hindered advances in the areas of control of cell proliferation and cancer, and to offer a way out of the impasse. Once we decided against writing an exhaustive review of all the data, the potential readers included all scientifically literate people, practicing physicians, with oncologists, pathologists and radiologists as their core, and of course, our fellow scientists and researchers. We understand that some may have wished we devoted more space to a detailed analysis of the data we refer to. We hope that they will understand the impossibility of doing this while remaining true to the stated aim of this book. In book chapters and research publications, we have offered and plan to continue offering a more detailed account of the different interpretations of data possible according to the opposing premises adopted by scientists. To those who would like to email us their supportive or dissenting views (societyofcells@tufts.edu), we will be posting and answering readers comments on the BIOS web site (http://www.bios.co.uk). In addition, we hope that biologists, philosophers, historians, and sociologists interested in these subjects may be stirred, if only to correct the shortcomings of our own analysis.

We have decided to restrict our discussion to two main propositions: the first is that *proliferation* is an built-in property of all cells, and second, that cancer is a disease that originally occurs at the tissue, rather than the subcellular, level of biological complexity.

## The problem of control of cell proliferation

The question, *Why do cells proliferate?* has not received the attention we feel it deserves despite its obvious relevance to the entirety of biology. On the one hand, unicellular organisms perform all the functions needed for survival and reproduction. As long as sufficient nutrients are available, these organisms proliferate, which is, for them, equivalent to reproduction; thus, their default state is *proliferation*. On the other hand, with the advent of multicellularity, cell proliferation and reproduction ceased to have a univocal relationship. In addition, constituent cells acquired the ability to specialize causing them to perform fewer functions, but more efficiently. These organisms learned how to simultaneously coordinate and control the proliferation of their diverse cell types. The proliferation rate of cells within these organisms is not directly dependent on nutrient supply, and proliferative quiescence appears as a physiological option.

In Part I of this book, encompassing the first six chapters, we discuss the reasons why the *Zeitgeist* has been that *quiescence* is the default state of cells in multicellular organisms. This perception is at the origin of the search for growth factors and oncogenes which has filled the scientific literature over the last 30 years. We make a brief historical analysis, describe the methodological hurdles that researchers overcame to explore this subject, and point out the successes and the failures in this quest. After evaluating all these elements, we make a case for considering that the default state of *all* cells is *proliferation*. Moreover, proliferation is a basic property of life, still needed for the propagation of the newly evolved organism. Hence, these multicelled organisms must have developed signals to thwart the constitutive ability of each of their cells to proliferate. In this context, control of cell replication becomes *negative control*.

## The problem of carcinogenesis

Since 1914, the prevalent theory of carcinogenesis, the *somatic mutation theory*, covered just one aspect of neoplasia, this is, the ability of these cells to breed true to type. The complex process whereby a mutation ends up forming the tissue dislocation typical of the carcinogenic lesion is not addressed by this theory nor its subordinate variants. To remedy this situation, we revive a theory of carcinogenesis that proposes that neoplasia is a defect of tissue organization, rather than a subcellular phenomenon as postulated by the somatic mutation theory. We call this the *tissue organization field theory*. Analyzed in three chapters in Part II, this theory offers a novel, comprehensive strategy to integrate data already collected and suggests new leads in understanding this disease.

For almost half a century, since Watson and Crick resolved the structure of the DNA molecule, the knowledge acquired at the molecular and intracellular levels of biology has been nothing short of astonishing. This discovery explained the physical bases for transmission genetics. The molecular biology dogma made accessible the connection between genetic information and its materialization into the proteins coded by these genes. Dimensions collapsed, and linear thought seemed to be a tool sufficient to pry open all the mysteries of organisms. François Jacob, in his book *The Possible and the Actual*, warned us about the limitations of linear thought in understanding complex, highly interactive phenomena comprised of heterogeneous components, such as different cell types and tissues. In short, what is called, emergent phenomena. However, for most scientists, the rapid pace of incremental knowledge in the first two decades of the molecular biology revolution led to the perception that this strategy recognized no limit and that eventually it would resolve problems grounded on higher levels of biological complexity.

In analyzing the contributions of molecular biology to the understanding of cancer, John Cairns recognized that the power of the molecular biology revolution resided in the switch from three dimensions to one dimension, which 'gave us forty years of increasing clarity'. 'But we are getting back to three dimensions and uncertainty returns'. '[We] now feel that at least we understand the principles and it is just the intricate details that defeat us'. To us, instead, this uncertainty indicates that we have misplaced the level of inquiry, which should have been at the tissue level, rather than at the cellular and subcellular levels. We will argue that the controls of cell proliferation and cancer are subjects grounded on the boundaries between the cellular and tissue levels of hierarchical complexities, and therefore, they require a conceptual and methodological approach consistent with their nature. In addition, we propose some avenues of inquiry that should significantly improve the outlook for understanding both the control of cell proliferation and of cancer.

In the last few years, researchers and lay commentators have wondered about whether or not the significant investment that has been devoted to understanding the carcinogenic process and to finding 'cures' for cancer was money wisely spent. In other words, critics expressed their concern about whether or not the basic assumptions under which these efforts were conducted were sound. Is there a justification for why these learned critics pose these questions? We think that the answer is yes. In fact, we believe that the scientific community and those who support their effort should welcome this challenge. The evidence shows that the current prevalent paradigms to explain cell proliferation and cancer are full of contradictions and are collapsing despite all the attention and resources committed to them. Our summarized message is this . . . Come to terms with the frustrating experience learned in the last three decades, let's go back to the drawing board, and start again!

# Acknowledgements

The origins of this book go back to 1976, when our epiphany about the exclusive negative control of cell proliferation took place. We were then conducting research at the INSERM Unité 34 in Lyon, France. We thank José Saez and the members of the Unité 34 who hosted us warmly and provided the *laissez-faire* environment that we needed. At the end of 1977, we became convinced that, in contrast to the expected outcome of our dated working hypothesis, estradiol, the female sex hormone, did not directly affect the proliferation rate of estrogen target cells. We then concluded that the data on the control of cell proliferation would fit better into a biological approach grounded on evolutionary principles. Christian Laugier, from the INSA-Lyon, helped us to flesh out our first approximation of the negative control hypothesis. His intellect and determination to test our hypothesis were critical to its initial development. Our ideas about carcinogenesis, instead, took shape recently, while writing the final drafts of this book.

We have tried to do justice to those who participated in the various stages of our journey regardless of which side of the issues they were in. Doing and searching for justice is probably an impossible task. We can only express our aims and plead guilty to any involuntary omission that failed to give credit where credit was due.

Several colleagues helped in reviewing one or more chapters of this book. Their suggestions helped in clarifying complex issues. However, we take responsibility for the final version we are presenting. Special thanks go to our colleagues at Tufts, Sheldon Krimsky, Ron DeLellis, and Elena Massarotti, to our colleague and friend Nicolas Olea, and to three external reviewers who wished to remain anonymous. They made sensible, incisive suggestions regarding content and style.

We are fully aware that, regardless of the merits of the paradigm switch which we have been proposing for over two decades, the vagaries of the funding process could have prevented us from pursuing our research. We are grateful to Study Section members from the National Institutes of Health, National Science Foundation and private foundations who reviewed our research applications. We extend our gratitude to the Branch Directors and Program Directors of the National Cancer Institute (especially, Alan Rabson, Colette Freeman, Brian Kimes, Suresh Mohla, and Barbara Bynum), the National Institute of Aging (David Finkelstein), and the National Science Foundation (Ed Berger, Eve Barak, Elvira Doman, and Stacia Sower). In a system that is naturally averse to risky research programs, they made our task at the bench easier. Our gratitude to them will never be sufficient.

The writing effort was made less painful by the French and Spanish Governments through the Université de Paris-Denis Diderot and the Universidad de Granada, respectively, where we spent our sabbatical leaves. Special thanks are due to Emmanuel A. Nunez, from the Hopital X. Bichat-Claude Bernard, in Paris, who made this task possible by his fine hosting and superb intellect. Nicolas and Fatima Olea, friends and colleagues, warmly hosted us as well in Granada. We thank Howard Bern, John McLachlan, Jack Gorsky, George Stancel, Sean Carroll, Elwood Jensen, Elisabeth Gateff, Harry Rubin, Jack Little, Gershom Zajicek, Pierre Chambon, François Jacob, Scott Gilbert, Richard Lewontin, Alfred Tauber, David Kirk, Juan Rosai, and the late Olav H. Iversen and Wallace H. Clark, Jr. for the stimulating discussions we had with them at different periods during our exciting journey.

We thank Tufts University School of Medicine for its culture of 'benign neglect'. Among several Tufts colleagues who helped us during this trek, we wish to identify for especial recognition Joseph J. Byrne, Larry Cavazos, Jerry Smulow, Karen Hitchcock, Murray Blair, Jim Liddy, Dan Fitzpatrick, and Charlie Ceurvels. Many people that passed through our laboratories deserve a special place in our appreciation: without them, our task would have been extremely improbable. Among those who deserve especial mention are Bob Schatz, Renee Silvia, Jim Bass, John Papendorp, Jozsef Szelei, Peter Geck, Cheryl Michaelson, Maria Luizzi, Anita Oles, Nancy Prechtl, Marcelle Desronvil, and Beau Weill who have provided editorial input that helped in making this book easier to digest. In addition, throughout the years, family members, colleagues and friends encouraged us, had faith in us, challenged us, ridiculed us, ignored us, and yet made this journey possible. They know who they are. They all helped us reach our goals in their own idiosyncratic way. Thanks to all.

## Chapter 1

# Cell Proliferation: The Background and the Premises

"How odd it is that anyone should not see that all observation should be for or against some view if it is to be of any service."

*Charles Darwin, quoted in: E. Mayr (1997) This Is Biology. Harvard University Press, p. 25*

"... Although existing hypothesis may be accused of being too specialized, too vague, not pertinent, or even inaccurate, hypotheses are made to be criticized, and each in its own way serves to sharpen our perspective ..."

*Richard J. Goss (1965) Adaptive Growth. Academic Press, New York, p. 42*

## Introduction

The biological world of which we are a part is a result of the huge numbers of cell replications since life began. Self-replication is the most fundamental property of all living organisms and cell proliferation is at the core of this process. Often, the notion of cell proliferation in multicellular organisms has been confused with that of 'growth'. The control of cell proliferation has received comparatively scant attention and the reasons for this will be discussed. We have selected the 1950s as a critical period in this historical account which originally began about a century ago. The resolution of the chemical structure of the DNA molecule caused many researchers to switch their views on how the control of cell proliferation and cancer were viewed thereafter. In addition, several key events (methodological as well as conceptual) related to the subjects of this book also took place during this decade, and we will refer to them as the narrative evolves. The publication of Paul Weiss' review in 1960 [1] marked the end of a period where physiologists, pathologists and biophysicists dealt exhaustively with the subject of organ and organismic growth. These researchers developed mathematical models that would accommodate the data collected during the first half of this century, for example, when describing the increase in size and weight of organs and organisms as a function of time during development and organ regeneration. After 1960, cell biologists and molecular biologists moved into the field of growth and proliferation with the overt intention of providing the 'mechanistic' understanding of growth. Prior to 1960, there had been little analysis of the relationship between cell proliferation and growth in tissues, organs or organisms. We are now in the midst of a period where knowledge accumulated at individual hierarchical levels is being integrated across several hierarchical levels. This has proven to be not an easy process.

## A brief history

Scientific controversies are seldom resolved by new results. Often, it is the re-interpretation of results into a more cogent theory that promises to solve controversies. An historical account of concepts about cell proliferation and growth reveals that change occurred mostly when old notions were dropped, rather than when new results were published. It is sobering to take a candid look at the original literature and try to detach oneself from concepts taken for granted today. A case should be made in favor of looking at competing hypotheses in the context of their times, for even the framing of the questions asked is firmly grounded in and therefore dependent on the *Zeitgeist*. The design of experiments also depends on the hypotheses they are supposed to explore. Most biology textbooks tell a tale of linear progress that rarely reflects the reality of the scientific enterprise.

### *Cell theory*

Dutrochet claimed in 1824 that cells were the building blocks of life in both animals and plants. He did not elaborate on the theoretical benefits that such a concept would extend to the understanding of biology. According to Henry Harris, Dumortier, in 1832, was the first to demonstrate cell division in multicellular organisms and attributed to this phenomenon the correct significance in a biological context, that is, that cells only originated from pre-existing cells [2]. However, it is generally recognized that the father of the cell theory was Schwann, who suggested in 1839 that cells both in animals and plants were their building blocks, explicitly locating all the functions of living organisms within these units. He divided these activities into 'plastic' (i.e. pertaining to growth) and 'metabolic' (i.e. responsible for chemical changes such as respiration). Interestingly, Schwann thought that cells were generated by a process akin to crystal formation, whereby new cells were formed extracellularly and accrued to other cells to form tissues. This opposed the existing view of Schleiden who proposed that cells were generated from within cells, also by a process akin to crystallization. Even after Remak described the accrual of new cells by direct cell division of the frog egg, and labeled the crystallization theory as 'apocryphal', the Schwann–Schleiden theory remained dominant. At this particular regard, Bechtel [3] suggests that rejection of a theory does not rely on being contradicted by experimental data. Other factors, such as the theoretical implications of the novel concept play a more important role. In this case, it was the prevailing desire of those in the physiological community to relate all biological phenomena to chemically based mechanisms that sustained Schwann's theory, despite evidence to the contrary.

By the 1850s and 60s, emphasis changed to matters related to the protoplasm and cell division. Relying on work by Remak, Virchow, a pathologist, declared that the cell is the fundamental unit in physiology and pathology. Henceforth, cells were believed to be generated from pre-existing cells by direct cell division. In turn, Bernard introduced the notion that cells should communicate with each other in an internal environment that must be maintained quasi constant. Eventually, there was a significant compromise that allowed for cell proliferation to be explained by direct cell division without committing to a physicochemical explanation of the process [3, 4].

### *Mitosis*

The cell theory established that the cell is the structural unit of life in both plants and animals. In 1855, Virchow proposed that cells were also units of reproduction of living matter. Thus, not until the cell theory was accepted could the question 'why do cells proliferate?' be asked. Little was known then of how cells divided. The role of the nucleus was obscure because its outline disappeared during mitosis. Several biologists described metaphase plates during the first half of the 19th century, but it was not until the 1870s that mitosis was considered as a part of the mechanism of cell division. Roux produced the most interesting interpretation of why a complex process such as mitosis took place when, at first sight, simple cell division might have sufficed. He proposed that since the nucleus was heterogeneous, mitosis was the mechanism whereby half of each different cellular constituent was passed to each daughter cell. While what are known today as 'transmission geneticists' were interested in discovering mechanisms that warranted the division of nuclear material in two exact copies, 'developmental biologists' (embryologists) wanted to find out how cell diversity was generated from a zygote, and saw cell division as a means to generate these unequal phenotypes [4]. All along, embryologists were concerned with the study of determination and 'differentiation'. Thus, while this was an acceptable concept to explain cellular diversity, it created a new problem: did somatic cells have a complete set of genes? The answer to this question had to await many changes of opinion, as well as the generation of new data through hypotheses which established the conservation of the full genome in somatic cells while each cell type expressed a different repertory of genes. This was settled when nuclei were transplanted into enucleated eggs and whole individuals developed [5, 6].

Over the past 100 years, many misconceptions regarding cell proliferation have remained as folklore, such as the view that 'undifferentiated' cells proliferate, while 'differentiated' ones do not. The aim of this book is to uncover these misconceptions and offer alternative hypotheses and conclusions. Not all of these may be correct, but by questioning the current dogma it is hoped that the subject may be viewed with a new openness.

### *Organismal growth. Cell proliferation as 'growth'*

Throughout the 20th century, biologists reasoned that the number of cells in a given organ depended on the rates of cell proliferation and of cell loss. They described how cell volume was related to ploidy and/or the presence of several nuclei. The control of cell proliferation was investigated only in a peripheral way.

Why did the control of cell proliferation fail to become a subject of study for those that subscribed to the cell theory? Several reasons may be offered to explain a lack of interest on the subject of control of cell proliferation. On the one hand, techniques to measure cell proliferation rates at the tissue level by detecting mitotic cells were laborious and inaccurate. Until the introduction of labeled thymidine in the 1950s, it was impossible to distinguish between cells that had recently divided from others that were quiescent. On the other hand, a search for agents that could modify the rate of cell proliferation in a tissue of interest had to be found before the subject of control of cell proliferation could be examined experimentally.

Manipulations that resulted in the increase of proliferative rates in the tissues of adult animals were equated with a 'positive' signal (i.e. a stimulus or a trigger). The perturbing agent may have had physiological relevance, such as estrogen-mediated proliferation of the uterine lining, or liver regeneration after partial hepatectomy. Often, toxic agents, such as lead or mercury, were used to induce mitotic figures in kidney tubules. In other cases, a pharmacological agent, such as isoproterenol, 'induced' DNA synthesis in glandular epithelial cells of the salivary glands of mice and rats [7]. Regardless of the nature of the agents used, and their physiological relevance, researchers must have considered these agents as stimulatory; implicitly, they were validating the notion that the default state of cells in multicellular organisms was *quiescence* because cells proliferated in response to what they perceived as a direct stimulus.

The phenomenon of organ regeneration either after removal of a portion (e.g. the liver) or of one of two (e.g. the kidneys) was described at the end of the 19th century. These reports were unclear regarding the specific questions being asked from these experimental protocols. The authors' immediate interest centered around 'how' and 'what' questions (such as, is the underlying cause of growth hypertrophy or hyperplasia? Does nutrition play a central role in the regenerative process? How does regeneration happen? What biochemical parameters change during the regenerative process?). Questions like, how is cell proliferation controlled? or, why do these cells proliferate? were never stated nor addressed by these researchers.

Among those concerned with organismal growth, Paul Weiss observed that it must be controlled negatively. Weiss cautioned that the simplicity of the tools to measure organismal growth masked the complexity of the phenomenon being studied. Growth was not a single entity but a sum of diverse phenomena. The indiscriminate use of the word 'growth' has impeded to a large extent a rigorous treatment of the concept of cell proliferation as a discrete topic. Weiss showed his frustration when defining growth by stating that, 'To put it bluntly, "growth" is a term as vague, ambiguous and fuzzy as everyday language has ever produced' and he went on to include 'reproduction, increase in dimension, linear increase, gain in weight, gain in organic mass, cell multiplication, mitosis, cell migration, protein synthesis, and perhaps more' as part of what biologists consider acceptable definitions of 'growth'. To remedy these ambiguities, he proposed to 'itemize the accounting' and define as exactly as possible the specific parameter being considered [1].

Next, Weiss offered, if not a comprehensive solution to the problem of representing growth, at least a realistic evaluation of a state of affairs that has dogged the treatment of growth up to the present day. 'This (complexity) confront us with a major task, which we mostly dodge rather than face: dodge by attaching real biological meaning to such purely descriptive terms as "growth stimulation" or "depression"; rather than face by finding out just *why* a given system, subject to certain agents, turns out to be either larger or smaller than a chosen reference standard. For this is all we usually determine; and yet, we imply far more'.

Weiss softened his 'critical and somewhat defeatist' analysis by proposing that 'Given optimal conditions, superabundance of all prerequisites and freedom from active inhibitions, each cell strain, tissue and organ grows at its own characteristic rate'. He concluded squarely that given his own experimental results *'One might question whether the search for over-all "growth stimulators" is at all realistic and promising'* (our italics). Unaware of this statement, starting in 1977, we independently developed a research program under the premise that cell proliferation is a constitutive, dominant property of all cells [8–10].

Separately, Weiss and Kavanau [11] and von Bertalanffy [12] adopted the notion that the regulation of organ and organismic growth is under negative control. They also discussed cell proliferation in relation to growth as one of several important components, but they cautioned that other phenomena also contribute to growth in size, such as hypertrophy and accrual of extracellular substance. In other words, they were dealing within an organismal hierarchy and cell proliferation was not considered a reliable parameter in the quantitative evaluation of growth. Weiss suggested that in order to understand organismal growth it was *preferable* to disregard the organization of tissues in cellular and extracellular components. Instead, he proposed to group living matter according to their generative capacity into two categories:

- the **generative mass** was always intracellular and comprised the machinery for cell reproduction as well as the machinery to produce extracellular and intracellular components such as collagen, myofibrils, etc.;
- the **differentiated mass** was comprised of both the intracellular and extracellular components of a tissue that did not possess the ability to reproduce.

Weiss hypothesized that each cell type produced its own growth inhibitor, a soluble, secreted component whose plasma concentration was proportional to the 'protoplasm mass'. Loss of mass of that tissue was followed by regeneration because the plasma levels of inhibitor decreased as a consequence of loss of parenchyma; this explained why hepatocyte proliferation stopped when the liver mass equaled that of the intact mass present before the operation. Weiss' model called to attention plausible mechanisms in the growth process by formulating them mathematically. These underlying mechanisms are negative feedback loops.

We will define 'growth' from a hierarchical perspective, namely, 'growth' is an increase in size of an organ or organism. When dealing with the tissue hierarchical perspective, we can avoid the use of the notion of 'growth' by using the more accurate terms 'hyperplasia' for an increase in cell number, and 'hypertrophy' for an increase in cell size.

### *Cells in culture*

Cell and tissue culture techniques were developed because whole animal studies are seldom conclusive for demonstrating ultimate causation. Roux first explanted tissues in culture dishes in 1885. However, technical obstacles prevented the study of the control of cell proliferation. The advent of trypsinization made it possible in the 1950s to count cells to measure proliferation rates accurately. Also, gaps in knowledge over proximate causality on control of cell proliferation precluded asking the question, why do cells proliferate in multicellular organisms? at least until the 1950s. For instance, it was not obvious which could be a relevant physiological effector of cell proliferation until the 1930s, when sex steroids were purified. The other suspected effector, growth hormone, became sufficiently purified in the 1940s. In addition to these technical and conceptual limitations, another shortcoming in evaluating data fairly had to be dealt with: this was the anthropocentric factor. Proud of their achievement in maintaining and propagating cells, implicitly, experimentalists assumed that cells proliferated in culture conditions because *they* themselves provided the 'stimulus' present within the widely used embryo extracts containing 'growth factors'. Originally, growth factors were meant to be the substances and conditions (pH, oxygen pressure, etc.) necessary for the optimal growth of microorganisms, rather than *signals* that controlled their proliferation.

## Understanding the control of cell proliferation

### *The cell cycle*

From the identification of mitosis as the process of nuclear division to the establishment of the phases of the cell cycle in the 1950s, a period of about 70 years elapsed. Nucleic acids were discovered by Miescher in the 1860s and, later on, it became clear that they were the main components of chromatin, and therefore, of chromosomes. The role of chromosomes on heredity was stated clearly by Sutton and Boveri in the first decade of the 20th century [13]. The work of Morgan, at Columbia University, his associates, and successors, clarified most of the problems of transmission genetics and, by establishing a chromosome map, provided physical evidence that the genetic material resided in chromosomes. Avery showed—as conclusively as it could ever be shown—that the genetic material was DNA, in spite of the doubts of some powerful skeptics [14]. In turn, Watson and Crick established the complementary nature of DNA, that is, the structural bases for genome replication [15]. A technological advance, the development of radioactive tracers in the 1940s, permitted the study of DNA synthesis, and thus, to assess when DNA was replicated along the cycle; this led to the description of the cell cycle as we know it today. Howard and Pelc in 1953 described the cell cycle in the roots of fava beans (*Vicia fava*) [16]; this generated a wave of research on cell cycle kinetics in normal and cancer tissues and mammalian cells in culture. The cell cycle was then defined by two measurable events, DNA synthesis (S phase) and mitosis (M phase), and two 'silent' intervals, called $G_1$ phase (the period between completion of mitosis and the start of DNA synthesis), and $G_2$ phase (the period between the end of DNA synthesis and the beginning of mitosis) (*Figure 1.1*).

### *Control of cell proliferation versus control of the cell cycle*

The concept of proliferative quiescence was interpreted as being a separate state from the $G_1$ phase of the cycle by some, and as a prolonged $G_1$ phase by others. This led to two views about regulation of cell proliferation. Through the first view, true quiescence only exists in fixed post-mitotic cells. All other apparently quiescent cells would be performing the cell cycle, but very slowly. In this case, control of cell proliferation becomes synonymous with control of $G_1$ traverse. The latter concept led to the belief that dissecting the 'molecular' steps that constitute the cell cycle will necessarily shed light on the mechanisms that control cell proliferation. The tendency among those favoring this latter hypothesis was to look for some key marker within the $G_1$ phase that would consistently predict sure entrance into the S phase [17]. An alternative view is that those who believed in a truly quiescent state posited that regulation should be exerted by means of signals that switched cells in and out of the cycle. In this conceptual

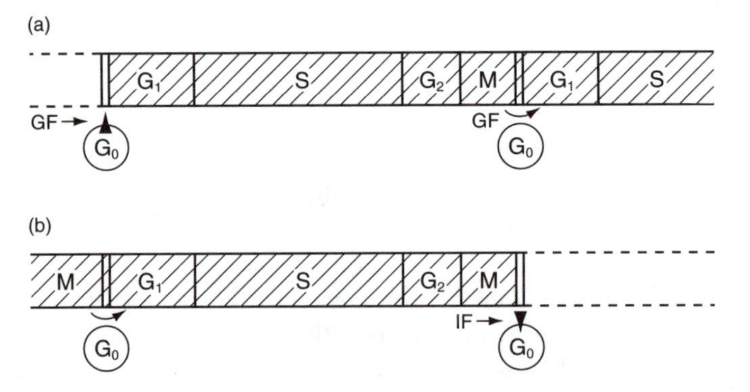


**Figure 1.1.** Schematic representation of the control of cell proliferation. (a) In the positive control hypothesis, the growth factor (GF) induces a quiescent ($G_0$) cell to enter the cycle. (b) In the negative control hypothesis, the inhibitory factor (IF) induces the cell to enter the $G_0$ state. The shaded area represents the cell undergoing the cycle. The arrowheads indicate the transition between the default and the regulated state.

frame, cell cycle becomes an automatic program of making two cells out of one. That is, when cells are committed to cycle they proceed traversing it inexorably. From this it follows that however interesting, studying the molecular events of the cell cycle is irrelevant to the cell's decision to proliferate or to become quiescent, for all cell types (normal and neoplastic) execute the same complex program to produce two cells out of one.

### *The nature of the default state: the crux of the matter*

The first conceptual decision to be made is to subscribe to the proposition that cells are either always performing the cell cycle, or they can move in and out of it undergoing cell proliferation arrest, a condition operationally defined as $G_0$. If cells move in and out of the cycle, one of these two states represents a **default**. If the default state is *quiescence*, proliferation is necessarily controlled by positive signals, a task assigned to the many 'growth factors' described in the last 40 years. This option is called the positive control hypothesis (*Figure 1.1*). If, to the contrary, the default state is *proliferation*, control is necessarily mediated by negative signals, represented by 'inhibitory factors'. This option is called the negative control hypothesis (*Figure 1.1*). Alternatively, if cells remain in the cycle permanently, control of cell proliferation and control of the cell cycle traverse are one and the same, and therefore, the default state is *proliferation*.

These views lead to further conceptual differences. While in unicellular organisms, *proliferation* is the acknowledged default state, in multicellular organisms the default state of cells is admittedly more difficult to assess because the internal milieu contains not only nutrients, but also signals to control the proliferative activity of many cell types. This means that to study control of cell proliferation experimentally the researcher unavoidably has to decide *a priori* whether the default state is *quiescence* or *proliferation*. Clearly, these options are mutually exclusive.

## The hierarchical organization of nature

Relevant to the analysis of control of cell proliferation is the perennial controversy of organicism versus reductionism. It deserves attention because biologists study control of cell proliferation ultimately to understand how multicellular organisms evolved from a single cell or a zygote (individual history), and need to ask what is the evolutionary history of this most basic process. Occasionally, by focusing our attention at lower hierarchical levels (biochemical and molecular) we lose perspective of the very reason of our research on multicellular beings, that reason being to understand organisms, including ourselves.

The success of biochemistry was based on the premise that the interaction of discrete components in a test tube would give clear, unambiguous answers because of the limited number of variables affecting the outcome of those interactions. Limiting the number of components may be a convenient strategy to identify causal agents; however, this should not quell concerns that there are other unsuspected, or ignored, variables that can significantly affect the results obtained with a limited strategy. This is in fact what may be happening in real life; hence, the lack of reproducibility of results that once worked in the test tube but fail to occur in more complex, organized environments (a cell, a tissue, an organ, a system, an organism).

From an evolutionary perspective, biology is the history of emergent phenomena and chance. From simple molecules emerged more complex molecules capable of self-replication. Dennett views life evolving as algorithms which produce new algorithms [18]. These mindless algorithms explain design from the bottom up, thus dispensing with vitalist arguments about the creation of life. Entities at one level of organization interact giving rise to a more complex level of organization, such as cells organizing themselves into tissues, these into organs, these into multi-organ systems, and these into organisms. As we move 'up' along levels of complexity, we find that new, unpredictable properties emerge (**emergence**) [19].

For most biological phenomena, exploring levels of complexity lower than that at which the phenomenon is observed usually adds little to what was learned at the original level of inquiry. For example, there is no need to understand the structure of the muscle fiber component myosin to explain how the heart works as a pump. In fact, people have been kept alive when their damaged heart was replaced by a man-made mechanical pump containing absolutely no myosin. On the other hand, lower level analysis sometimes reveals a continuum of features that enrich the understanding of the whole. For example, the mechanism of striated muscle contraction could only be understood when its subcellular structure was revealed. This implies that to comprehend a given biological phenomenon each hierarchical level should be studied without expecting that lower levels of inquiry will contribute to the understanding of that particular phenomenon. In addition to 'bottom-up' emergence, there is a reverse emergence, whereby the organism as a whole affects the properties of its parts [19].

The hierarchical organization of multicellular organisms is beyond dispute. However, once the existence of emergent phenomena is acknowledged, we are left with no guidance about how to explore them systematically. On the one hand, we tend to use analytical reductionism to study (deconstruct?) biological problems; on the other hand, we may destroy emergent phenomena applying this experimental approach. One way to detect these emergent phenomena in the study of cell proliferation control is by applying the *in animal–in culture* approach (see Chapter 5). Clearly, however, methodology to study emergent phenomena is sorely lacking.

## Conclusions

In the course of this first chapter, we briefly reviewed the history of the cell theory and its implications on epistemological arguments on the control of cell proliferation in multicellular organisms. An important lesson learned from the history of biology is that established concepts (cell theory, gene, etc.) and the original supportive experimental data on which they are based, were inconclusive when they were originally presented. For example, had Mendel chosen a plant other than peas (*Pisum sativum*) for his now famous experiments, it is likely that he would have obtained results impossible to interpret due to crossing-over, linkage effects, additive characters or polyploidy. Were Mendel's conclusions less valid because they could not be reproduced easily in other species? In hindsight, he chose the most adequate model, for only after Mendel's rules were accepted, could phenomena like linkage and crossing-over be understood. This is just one example to illustrate that the very complexity of nature may mislead us when interpreting experimental results. In an experiment there are a lot more (unknown) variables than the elements relevant to the hypothesis being explored.

In studying any natural phenomenon, taking choices among the premises to be adopted becomes unavoidable. Pragmatically, this is the only sound path to follow. The body of data available to the researcher always shows inconsistencies, contradictions and exceptions. Choosing to trust one set of data over another, or to adopt one premise over others, is subject to a reasoned, though sometimes intuitive, decision. At the beginning of the 'Scientific Revolution', Kepler used only Tycho Brahe's data on the position of the planets to describe their motion around the sun, and found their orbits to be elliptical. Had he considered all available data, he would have found it impossible to fit them into any coherent model. In the same vein, referring to Galileo's Law concerning falling bodies, his disciple, Torricelli said: 'If balls of lead, iron and stone do not obey the law, so much worse for them; then we say that we do not speak of them' [12]. This means, as long as the concept makes sense, results that do not fit may be construed to indicate complicating factors. Of course, they may also indicate that the theory is wrong, but this will not be known for a while. Therefore, strictly intuitive choices occasionally have to be made, even at the risk of being wrong. Someone may eventually falsify them and move on to more promising theories. After all, uncertainty is the daily concern of scientists.

So we have made our choices. In the following chapters, we present arguments for having switched premises and, consequently, working hypotheses, and we back up our decisions in an increasingly complex fashion.

## References


1. **Weiss, P.** (1960) What is growth? In: *Fundamental Aspects of Normal and Malignant Growth* (ed. W.W. Nowinski). Elsevier Publishing Co., Amsterdam, pp. 1–16.
2. **Harris, H.** (1995) *The Cells of the Body: A History of Somatic Cell Genetics*. Cold Spring Harbor Laboratory Press, Plainview, NY, pp. 1–30.
3. **Bechtel, W.** (1984) The evolution of our understanding of the cell: a study in the dynamics of scientific progress. *Stud. Hist. Phil. Sci.* **15**: 309–356.
4. **Mayr, E.** (1982) *The Growth of Biological Thought: Diversity, Evolution, and Inheritance*. Belknap Press, Cambridge, MA, pp. 652–680.
5. **Briggs, R. and King, T.J.** (1952) Transplantation of living nuclei from blastula cells into enucleated frogs' eggs. *Proc. Natl Acad. Sci. USA* **38**: 455–463.
6. **Gurdon, J.B.** (1968) Transplanted nuclei and cell differentiation. *Sci. American* **219**: 24–35.
7. **Baserga, R.** (1976) *Multiplication and Division of Mammalian Cells*. M. Dekker, New York, pp. 58–65.
8. **Soto, A.M. and Sonnenschein, C.** (1984) Mechanism of estrogen action on cellular proliferation: evidence for indirect and negative control on cloned breast tumor cells. *Biochem. Biophys. Res. Commun.* **122**: 1097–1103.
9. **Soto, A.M. and Sonnenschein, C.** (1987) Cell proliferation of estrogen-sensitive cells: the case for negative control. *Endocr. Rev.* **8**: 44–52.
10. **Sonnenschein, C. and Soto, A.M.** (1991) Cell proliferation in metazoans: negative control mechanisms. In: *Regulatory Mechanisms in Breast Cancer* (eds M.E. Lippman and R.B. Dickson). Kluwer, Boston, MA, pp. 171–194.
11. **Weiss, P. and Kavanau, J.L.** (1957) A model of growth control in mathematical terms. *J. Gen. Physiol.* **41**: 1–47.
12. **von Bertalanffy, L.** (1960) Principles and theory of growth. In: *Fundamental Aspects of Normal and Malignant Growth* (ed. W.W. Nowinski). Elsevier Publishing Co., Amsterdam, pp. 137–259.

13. **Mayr, E.** (1982) *The Growth of Biological Thought: Diversity, Evolution, and Inheritance*. Belknap Press, Cambridge, MA, pp. 727–776.

14. **Judson, H.F.** (1995) *The Eighth Day of Creation*. Penguin Books, Toronto, ON, pp. 33–41.

15. **Watson, J.D. and Crick, F.H.C.** (1953) Molecular structure of nucleic acids: a structure for deoxyribose nucleic acid. *Nature* **171**: 737–738.

16. **Howard, A. and Pelc, S.R.** (1953) Synthesis of DNA in normal and irradiated cells and its relation to chromosome breakage. *Heredity (Suppl)* **6**: 261–273.

17. **Pardee, A.B.** (1974) A restriction point for control of normal animal cell proliferation. *Proc. Natl Acad. Sci. USA* **71**: 1286–1290.

18. **Dennett, D.C.** (1995) *Darwin's Dangerous Idea*. Simon & Schuster, New York, pp. 48–60.

19. **Mayr, E.** (1982) *The Growth of Biological Thought: Diversity, Evolution, and Inheritance*. Belknap Press, Cambridge, MA, pp. 1–146.

## Chapter 2

# Cell Proliferation, Cell Nutrition and Evolution

"What is a paradigm, Uncle Bob?
Dear Peterkin, when will you ever grow up? Nobody cares what a paradigm is. 'Paradigm' exists to be said, not thought about. It probably doesn't mean anything at all. Just say it now and then, and easily bullied people will probably think you know what you are talking about."

*Russell Baker (9/2/95) The New York Times*

"The fact that life is hierarchical and that lower units are conservative demands the interpretation that each unit persisting in life's hierarchy today must have become resistant to further modification."

*Leo W. Buss (1987) The Evolution of Individuality. Princeton University Press, Princeton, NJ, p. 188*

In this chapter, we explore the context under which the control of cell proliferation was studied in the last 100 years, and the question of why the default state for cells in metazoa has been assumed to be *quiescence*.

## Evolutionary perspective on the control of cell proliferation

### *From prokaryotes to unicellular eukaryotes*

François Jacob wrote that the dream of a bacterium is to become two [1]. This aphorism illustrates what these organisms are about. Prokaryotes have adapted to occupy very disparate niches, from thermophiles, living at high temperatures where no other organism could survive, to saprophytes and pathogens which need a host. Since the time of Pasteur, much research has been aimed at making bacteria thrive in laboratory settings. Culture media were developed empirically;

for example, microorganisms living on grass blades were grown in a hay infusion. By analogy, it was believed that meat broth would supply an adequate environment for human pathogens. Clinical microbiology was born on these principles.

In the last half of the 19th century, finding pathogens and demonstrating that they caused a given disease had occupied the attention of microbiologists. Later on, microorganisms were studied because they raised interesting biological questions. To understand their way of living, it was necessary to grow them in large volumes in a reproducible fashion. According to their nutritional needs, there are prokaryotes that are **autotrophs** (i.e. they thrive just on a source of nitrogen and carbon); and, there are those that need complex diets, called **auxotrophs**. Some pathogens cannot thrive in laboratory environments even in a rich 'defined' medium. While finding out the nutritional requirements of particular bacteria, scientists began referring to these requirements as **growth factors**. Experimental biologists discovered that the nutrient quantity and quality is what most influences the proliferation rate of bacteria. Proliferative quiescence is not an option for them; starvation usually results in cell death.

Other organisms evolved ways to survive during starvation. For example, if a culture of the bacterium *Serratia marcensis* is allowed to grow to the highest achievable density without replenishing nutrients, the bacteria will survive for prolonged periods in what is called **stationary phase**, whereby the number of living cells will oscillate around the top density value [2]. The explanation of this phenomenon is a sort of cannibalism: by secreting proteases that will digest the dead bacteria in the growth medium, live bacteria get nourishment. Other organisms have evolved the ability to generate spores, which are latent life capsules. These spores can withstand inhospitable environments, and will germinate when finding themselves in a more favorable situation, switching to a vegetative state. This type of suspended life is not comparable with quiescence, since spores have practically no metabolic activity. Moreover, the chances of any one spore regaining vegetative life are small. To summarize, proliferation is a constitutive property of prokaryotes [3, 4].

Like prokaryotes, unicellular eukaryotes may live independent lives, live as symbionts or as pathogens. Their *raison d'etre* is, again, to propagate. Some have chloroplasts such as *Euglena* and are autotrophs; those that do not have chloroplasts usually live by eating bacteria. Parasites can be so very specialized regarding their nutritional requirements that they are totally dependent on their hosts. Finally, some have developed the ability to reproduce sexually, forming gametes when their nutrient supply runs low. For example, some yeast undergo meiosis; the haploid gametes formed are of a specific mating type and undergo conjugation with the opposite mating type to form the diploid organism [5].

What is the advantage of these strategies over that of continuing asexual vegetative reproduction? Sexual reproduction results in a lower number of offspring, since one organism is formed out of two gamete cells, while vegetative reproduction produces two organisms from one. In evolutionary terms, a sound strategy for survival is to produce the highest possible number of descendants; however, when the food supply runs low, this 'buying time' strategy allows survival of fewer organisms, which is more tolerable than sure extinction. These examples illustrate that the ability to proliferate is constitutive in these organisms, and that control mechanisms evolved to curtail reproduction during starvation. Hence, quiescence has never been an option for unicellular organisms aiming to reach the limit of their proliferative capabilities.

### *From unicellular eukaryotes to metazoa and metaphyta*

From an evolutionary perspective, namely that of the continuity of life, multicellular organisms (metazoa and metaphyta) necessarily evolved from unicellular ones [6]. A more complex control of cell proliferation evolved in the multicellular organisms. Reproduction and cell proliferation ceased to be one and the same because in metazoa the germ plasm is segregated from the somatic components. Specialized cellular function and developmental constraints in the body plan require that the proliferative activity of somatic cells be restrained. Somatic cells, like the host they are part of, are eventually destined to die [7]. The proliferation of somatic cells is no longer directly dependent on the availability of nutrients; this is because cells and tissues take their nutrients from the internal milieu of quasi-constant composition. Cells are exposed to those nutrients regardless of whether they become engaged in proliferation or not. Moreover, the rate of proliferation varies among cell and tissue types and within a given organ at different developmental and physiological stages. Some cell types, like neurons in the central nervous system (CNS) or liver cells in adult mammals remain proliferatively arrested during prolonged periods of time, even when the CNS and the liver are bathed by nutrients (since the wall of sinusoids are discontinuous there is no doubt that nutrients are indeed available to hepatocytes).

#### *A stealth switch in the default state of multicellular organisms?*

What happened to the constitutive ability of cells to proliferate during the advent of multicellularity? The significant structural differences brought about by multicellularity must have given rise to a number of emergent properties in these organisms. Among these new properties, responding to specific signals (extracellular, but inside the organism) to regulate the rate of cell proliferation must have been crucial to these novel conditions. However, *proliferation* should have remained as the default state of cells in multicellular organisms. We could find no compelling reason to conclude otherwise.

When *proliferation* is accepted to be the default state, it follows that the control has to be negative. The contrary interpretation is that, somehow, with the advent of multicellularity *quiescence* has become the constitutive state. Since all cell cycle markers studied so far have been highly conserved from yeast to humans, it is difficult to envisage circumstances whereby the default state of *proliferation* inherent to unicellular organisms would have switched to *quiescence*.

Although it is seldom articulated, this is what is implied in the search for growth factors. This concept is explicitly described in the textbook *Molecular Biology of the Cell* [4]:

> "To produce and maintain the intricate organization of the body, the component cells must obey strict controls that limit their proliferation. At any instant most cells in the adult are not growing or dividing but instead are in a resting state, performing their specialized function while retired from the division cycle. Because nutrients are plentiful in the tissues of the body, the cells must refrain from proliferating in circumstances where a yeast or bacterium would proliferate readily. What accounts for the difference?
> We shall see that for the cells of a multicellular animal, nutrients are not enough: **in order to grow and divide, a cell must receive specific positive signals from other cells**. Many of these signals are protein *growth factors*, which bind to complementary receptors in the plasma membrane to stimulate cell proliferation. These positive signals act by overriding intracellular negative controls that otherwise restrain growth and block progress of the cell cycle control system. Thus, while a well-fed yeast cell proliferates unless it gets a negative signal (such as a mating factor) to halt, an animal cell halts unless it gets a positive signal to proliferate."

Although presented as a fact, this concept is just an opinion. The authors do not elaborate on how this evolutionary novelty may have come about. It should be pointed out that this is a widely held notion among active participants in this field who have spelled out the implications of adopting this premise:

> ". . . In the absence of growth factors (GFs), a normal cell will exit from its growth cycle, enter the resting or quiescent state known as $G_0$, and remain there for days, weeks or even years. When GFs are supplied, the cell will leave $G_0$ and reenter the active growth cycle. The presence or absence of GFs is thus the basic element that determines whether or not a cell will grow . . . [8]."

The concept of growth factors as signals to induce cell proliferation did not emerge from evolutionary theory. Instead, once growth factors were operationally defined, the need to integrate them into the core of cell biology required *ad hoc* explanations such as the one quoted above.

### *Reproduction and sex*

Unicellular organisms have evolved multiple strategies to reproduce. Some are asexual, others are sexual and many represent a combination of both. If the main function of every organism is to reproduce, why is it that sexual reproduction was not selected against? In other words, if asexual reproduction generates two individuals from one, what is the advantage of sexual reproduction, that in its simplest form uses energy to produce gametes, and then, from the fusion of two gametes, only one organism is formed? This seems a wasteful effort. As mentioned above, many unicellular organisms reproduce vegetatively (asexually) when nutrients are plentiful, and switch to sexual reproduction when starved. Sexual reproduction, through the process of crossing-over, produces variation among individuals, 'the material out of which natural selection produces new species' [7]. Hence, sexual reproduction has two consequences: it allows survival during starvation, and it produces variation.

This link between nutrient availability and mode of reproduction does not stop with unicellular organisms. Insects that live in an environment where food supply is scarce and hard to find, but once found is plentiful, have evolved interesting reproductive strategies. Organisms cannot adapt to this unpredictability, but they can increase the likelihood of having their genes passed on to the next generation, by reproducing madly while the food supply is plentiful. The example of cecidomyan gall midges taken from S.J. Gould's book, *Ever Since Darwin*, illustrates this strategy [9]. These insects reproduce sexually, laying eggs that undergo larval and pupal molts and finally emerge as flies. When flies discover a source of nutrients (a mushroom), their offspring respond to the superabundance of food by reproducing parthenogenetically as larvae or pupae. They do not lay eggs; instead, their oocytes develop as flightless organisms inside their mothers' bodies. This continues until the food supply is exhausted, then the flies switch to sexual reproduction. The mature flies will search for the next mushroom where this cycle will start again. As with unicellular organisms, food abundance results in asexual reproduction that produces numerous offspring rapidly, and starvation results in sexual reproduction which is slow and numerically less efficient. Simply put, life is reproduction.

*Control of cell number in metazoa*

Let us turn for a moment to the size of the organs in adult individuals in metazoa. Organ size is maintained fairly constant throughout a process of repair whereby the organism regulates the rate of cell replacement. In other words, cell numbers are maintained through a balance between cell proliferation and cell death. While in certain organs, such as the liver, the rate of both cell death and cell proliferation is relatively low, and most cells are quiescent, in other tissues like the endometrium in the uterus, cell proliferation and cell death occur in cycles. Still, in rapidly renewing tissues, such as the hemopoietic ones, the rates of cell proliferation and death are high; here, cell numbers are mostly regulated by altering the rate of cell death [10].

There are two kinds of cell death, necrosis and apoptosis. This is often explained using an anthropocentric metaphor; necrosis is seen as death by murder, and apoptosis, death by suicide. These two types of cell death are recognized by the microscopic observation of tissues taken from animals because these cells display characteristic morphological features. Necrosis occurs when tissues are deprived of oxygen and nutrients, like the process of shedding the endometrium's inner-most layers during menstruation. Apoptosis, also called 'programmed cell death', occurs in the absence of external injuries; otherwise normal cells trigger a process that results in their own death.

Embryologists first observed the phenomenon of programmed cell death as a component of morphogenesis. The phenomenon of apoptosis in adult tissues was first described in the 1970s [11]. During the last 25 years this phenomenon has been intensely studied at the biochemical level. Indeed, morphological changes seen with the light microscope are the reflection of biochemical changes taking place in those cells. The search for a subcellular, biochemical explanation of this cellular phenomenon led to the identification of concurrent effects such as DNA degradation, the induction of certain genes, etc. [12, 13].

Control of cell death is a phenomenon conceptually independent from that of the control of cell proliferation. This chapter deals with the control of cell proliferation rather than that of the control of cell number. Thus, control of cell death will not be discussed further.

*Control of cell proliferation in multicellular organisms*

Cells in multicellular organisms are arranged in tissues and organs as a result of the unfolding of the developmental program. This program is unveiled through interactions among cells. Topology and history shape the pattern of gene expression of each cell, and modulate their proliferative activity [14]. These interactions are mediated by diverse means. They encompass the proteins that are present in the plasma membranes and allow cell-to-cell recognition. For example, a juxtacrine interaction is one in which a cell expresses a ligand in its plasma membrane that is recognized by a receptor in the membrane of an adjacent cell. A paracrine signal is secreted by a cell and is recognized by its neighboring cells. Another type of signaling occurs through junctional complexes that bind cells together. For instance, communicating junctions allow the passage of small molecules from cell to cell. Anchoring junctions attach cells and their cytoskeletons to other cells or to the extracellular matrix. These interactions shape the tridimensional organization of tissues and the organs where they reside, and also influence the proliferation pattern of the cells in that tissue [15–17]. There are also organismal signals that are produced by one organ, and act upon target cells located in distant organs. These are called endocrine signals (*Table 2.1*).

In summary, many levels of organization play a direct or indirect role on the control of the cell number in a given tissue or organ. The immense complexity of these interactions invites the use of reductionistic approaches of analysis. In this strategy, a first step is to formulate hypotheses about the default state of metazoan cells. The next step is the search for causal agents. If the default state adopted is *quiescence*, it follows that positive signals will be searched. If the default state adopted is *proliferation* the researcher will seek negative signals.

## The role of nutrition on the control of cell proliferation

Biological problems should be interpreted in the context of evolution. We argued above that it makes evolutionary sense for the default state of *proliferation* to be extended from unicellular to multicellular organisms. Is this concept testable? In principle, placing cells from metazoa in a milieu containing an adequate nutrient supply should serve this purpose; they would proliferate if the default state was *proliferation*, or they would not if the default state was *quiescence*. However, there is a limiting factor that hinders this test in metazoa: current know-how does not provide an adequate supply of nutrients in defined culture medium for all

**Table 2.1.** Mediators of the control of cell proliferation in metazoa

|  |  |
|---|---|
| 1. Juxtacrine mediators | (plasma membrane proteins, a ligand in one cell is recognized by a receptor in a neighboring cell) |
| 2. Paracrine mediators | (soluble factors secreted by one cell are recognized by neighboring cells) |
| 3. Cell-to-cell interactions | (junctional complexes) |
| 4. Extracellular matrix receptors | (integrins) |
| 5. Endocrine mediators | (blood-borne hormones) |

animal cells. Instead, we know that most cells *in culture* are successfully propagated in medium supplemented with serum. The role of serum as purveyor either of nutrients or of 'signals' (the growth factors proposed by those that believe that the default state is *quiescence*) depends on the premises one adopts regarding the default state (see also Chapters 4–6). Obviously, lack of knowledge on the nutritive requirements for cell viability and proliferation hinders an experimental verification of the nature of the default state in metazoa. To the contrary, cells from most plants easily fulfill the requirement to proliferate in chemically defined medium; hence, unambiguously, the default state in metaphyta is *proliferation* [18].

### *Tissue and cell culture as means to study complex phenomena*

The use of *in-culture* models grew from the inadequacy of whole animal models to generate answers to questions about causality. After initial attempts by Roux in 1885 to study neural development using embryonal tissues, Loeb, in 1897, was able to maintain several types of tissues in culture conditions. Full credit, however, has been bestowed on Harrison, the outstanding developmental and cellular biologist, who a decade or so later, began using this tool in a systematic way. Solving technical issues had pressing urgency for those that believed in the potential of tissue culture to elucidate biological problems. In those early years, technical issues like medium composition and the long-term maintenance of cells were important. At that time, cells could not be transplanted *in culture* because there were no easy and mild protocols to separate them from each other and from the culture flask surface, nor could they be stored frozen. In addition, there were no antibiotics or laminar flow hoods to reduce the incidence of microbial contamination.

Willmer pointed out that in the 1920s three main topics were dominant: (i) '*growth stimulation*' (our italics), (ii) 'the then less popular problems of differentiation and cell behavior' and (iii) 'the provision of an adequate diet for pure strains of growing cells' [19]. The development of defined medium for the propagation of cells *in culture* was a central issue among researchers. Then, as now, while some cells grew and could be propagated in such media, most others required the presence of serum for optimal rates and yields. However, Willmer also called attention to the fact that, in contrast to animal cells, plant cells proliferated 'so long as they are given the necessary salts, water and light'. Finally, he underlined the complexity of the living matter and the multiple sources of errors introduced when cells removed from the intact animal (or plant) were studied in the 'reduced' environment of tissue culture conditions.

In several chapters of a book edited by Willmer, experts discussed the meaning of serum supplementation in cell and tissue cultures [20]. The design of nutritive media consisting of an ever-expanding lists of amino acids, vitamins, salts, hormones, nucleosides and other arbitrarily selected defined and patently undefined (tissue extract) components occupied considerable time and effort of researchers. Their aim was to find the elusive formula in which a majority of cell lines would thrive in culture flasks. The underlying assumption, the nutritional dogma, was that metazoa needed a source of energy (sugar), essential amino acids, fatty acids, minerals and vitamins. Everything else was assumed to be synthesized by the cells. However, only a small number of cell lines were able to adapt to proliferate under those nutritionally restricted conditions. Most cells died when exposed to chemically defined media. For the most part, these efforts were based on the notion that cells could be adapted slowly (in terms of months) to proliferate equally well in chemically defined media at rates comparable to those observed in serum-supplemented media. This did not materialize and with some exceptions limited to manufacturing practices in the biotechnology area, serum-supplemented medium remains the accepted universal nutrient solution for cells *in culture.* Thus, serum-borne macromolecules are needed for the survival, proliferation, propagation and maintenance of a good number of desirable specialized phenotypes of metazoan cells *in culture.*

A question was not raised in Willmer's book, namely, 'What is missing in the culture dish that allowed healthy cells to exercise efficiently their cryptic proliferative capability while they were quiescent in the intact plant or animal?', or alternatively, 'What was present in the culture dish and absent in the animal or plant that justified the relentless proliferation of some cells in glass and, later on, plastic flasks?' All along, the implicit premise adopted to study the control of cell proliferation in normal and neoplastic cells was that the default state of these cells was *quiescence.* The name of the game was to look for stimulators of cell proliferation.

### *Quantitating cell proliferation*

Two technical developments made the quantitation of cell proliferation and, hence, the analysis of its control, easier. In the early 1950s, Moscona re-introduced the use of trypsin originally proposed by Peyton Rous to detach cells from each other and from the glass surface in which they grew [21, 22]. This simple procedure facilitated not only the propagation of cells over time without the traumatic treatment they were subjected to before (mechanical disruption of monolayers with rubber policemen and energetic pipetting) but provided means to quantify them in counting chambers with improved accuracy. This was a significant contribution. The other methodological improvement that became available in the 1960s, but was incorporated routinely in the 1970s, was the introduction of the electronic particle (cells, in this case) counting machine by Coulter.

Finally, in 1953, the unveiling of the model of the double helix for the structure of DNA by Watson and Crick marked the beginning of the molecular biology revolution. However, we consider that this revolution was not fully felt in the broad field of multicellular organisms until the 1970s when a sizable number of scientists working in prokaryotes switched their attention to somatic cell genetics (funding was a significant enticement in the USA, when the War on Cancer program was enacted in 1971). It is in Baserga's book *Multiplication and Division of Mammalian Cells*, where the impact of the molecular biology revolution is reflected upon [23].

*The controversy about defined medium versus serum-supplemented medium*

Why develop chemically defined media when maximal cell proliferation rates are obtained mostly with the undefined, but most effective, heterologous serum-supplemented media? As mentioned by Weymouth, data published in the 1930s about nutrition of rats using a defined diet were interpreted as meaning that amino acids, vitamins, lipids, and minerals were all the necessary nutrients [24, 25]. Therefore, it was assumed that if these nutrients were adequate for a whole organism, they should have been sufficient for cells *in culture*.

Along these lines, Eagle and Piez [26] tried to find out whether plasma proteins, as proposed by Kent and Gey [27], were used by cells as a source of amino acids. In order to produce radiolabeled serum proteins, they administered radioactive amino acids to rabbits, whose serum was supplemented to basic culture medium. They reasoned that if the plasma proteins were used as a source of amino acids, cells cultured in this medium will incorporate the labeled amino acids into their own proteins. Because of the low radioactivity detected in those cells, they concluded that plasma proteins were not used as a source of amino acids. This conclusion was valid inasmuch as it confirmed previous data that showed that medium without amino acids supplemented to dialyzed serum did not support proliferation. Incidentally, neither did serum albumin, which represents about 60% of the total serum protein. However, these experiments did not, and could not, rule out whether or not serum played a nutritive role. For example, they did not rule out whether serum provided other building blocks, such as lipids, which are carried by plasma lipoproteins, or that a small subpopulation of plasma proteins was used as a source of a particular amino acid which may be unstable in the non-reducing atmosphere of the tissue culture flasks. Nevertheless, this influential paper was accepted by the scientific community as evidence that plasma proteins played no other role than that of carriers of signals (proliferation regulators/growth factors) that would have induced the entry of cells into the cycle. This publication marked a change in attitude and nomenclature. Before then, growth factor just meant any nutrient that when added to culture medium contributed significantly to make possible the propagation of a given cell, or to increase its proliferation rate. After Eagle and Piez, growth factor meant a *signal* to move metazoan cells from quiescence to proliferation. In short, this notion strengthened the perception invoked by predecessors and contemporary observers that *quiescence* was indeed the default state in metazoa.

At the end of the 1950s and early 1960s, a series of research currents converged toward the field of somatic cell genetics and cell culture. In addition to efforts stemming from those groups who over the previous two decades were concerned with the empirical formulation of cell culture media (Earle, Waymouth, Evans, Sanford, Eagle), the field was significantly influenced by the contributions of Levi-Montalcini and Cohen. In the 1950s, Levi-Montalcini claimed that nerve growth factor (NGF) increased the number of cytoplasmic processes (neurites) in sympathetic neurons in culture conditions [28]. Cohen's data were interpreted as implying that mouse submaxillary gland extracts contained and secreted growth factors [29]. When injected, these factors stimulated the keratinization of skin and the early eruption of teeth of newborn mice [30]. These data are analyzed in a wider context in Chapter 4.

### *The accumulation of incriminating 'evidence'*

The perception that growth factors stimulated cell proliferation was seized by Sato's group to develop the notion that the macromolecular components of serum were a mixture of nutrients, hormones and growth factors [31]. Again, this notion would only be valid if indeed *quiescence* was the default state of cells in metazoa. It should also be pointed out that cells exposed to conventional nutrient solutions (DME, F10, etc.) without serum do not become quiescent but instead they die. A program aimed at replacing serum by 'defined' growth factors was set under way.

Serum-free media were made of conventional nutrient solutions (F10, DME, etc.) supplemented with hormones and putative growth factors chosen with the aim of supporting optimal proliferation of a given cell type. In spite of all the effort devoted to this program, it is recognized that the propagation of most cell types and cell lines still requires serum-supplemented medium [32]. Only a few cell lines have been adapted to propagate indefinitely in a serumless medium. Serumless media have also been used in short-term experiments, and by the biotechnology industry, to facilitate the purification of bioactive molecules secreted into the culture medium. In short, the almost 40-year research program based on the premise that *quiescence* was the default state in metazoa has neither developed effective serumless formulations for the long-term propagation of cell lines, nor has it demonstrated that the role of serum is to provide ultimate *signals* to trigger cell proliferation.

It should be acknowledged, however, that this is still a controversial subject. For example, in the textbook *Molecular Biology of the Cell* [33], it is stated that:

> "Until the early 1970s tissue culture was something of a blend of science and witchcraft. Although tissue fluid clots were replaced by dishes of liquid media containing specified quantities of small molecules such as salts, glucose, amino acids, and vitamins, most media also included a poorly defined mixture of macromolecules in the form of horse serum or fetal calf serum or a crude extract made from chick embryos. Such media are still used today for most routine cell culture, but they make it difficult for the investigator to know which specific macromolecules a particular cell needs to have in the medium in order to thrive and function normally.
> This difficulty led to the development of various *serum-free, chemically defined media*. In addition to the usual small molecules, such defined media contain one or more specific proteins that most cells require in order to survive and proliferate in culture. These include **growth factors**, which stimulate cell proliferation, and *transferrin*, which carries iron into cells. Many of the extracellular protein signaling molecules essential for the survival, development, and proliferation of specific cell types have been discovered by studies in cell culture, and the search for new ones has been made very much easier by the availability of serum-free, chemically defined media."

In spite of the labeling of the use of serum supplements as 'witchcraft', the statement above concedes that serum is used routinely for cell propagation. Serum is awaiting an aggressive research effort aimed at unravelling nutrition at the cellular hierarchical level.

### *Senescence in a flask*

A phenomenon called cell senescence was described whereby somatic cells from a number of species, including humans, proliferate in culture conditions for only a limited number of cell divisions. In the 1960s, data suggested that the number of cell replications achieved by human cells in culture conditions decreased as the age of the donor increased [34]. These data favored the proposition that these *in-culture* experiments reflected a phenomenon similar to that experienced by organisms who aged. Others challenged this interpretation. Much research was conducted in this field. However, it is presently unknown whether cell senescence *in culture* has any relationship with the aging of the whole organism [35]. In fact, several lines of evidence suggest that this is an artifact of cell culture. For example, established cell lines, which are able to be propagated indefinitely may undergo 'senescence' and later die after being exposed to serumless medium for a few generations despite being returned to optimum serum-supplemented medium [36]. It is also puzzling to observe that the 40–60 generations limit verified when using human fetal fibroblasts is not reproduced when human leukocytes are used to start these cultures; their lifespan can be extended indefinitely under routine conditions [37]. Because the subject of cell senescence does not relate directly with that of control of cell proliferation we refer those interested in the subject to a critical review by Harry Rubin [38].

### *Nutrition and the proliferation of plant cells*

Plant biologists look at what animal cell biologists do with an interest that is seldom reciprocated. One of the missed opportunities resulting from this lack of shared experiences was the oversight by animal cell biologists of the provocative statements advanced by Steward regarding the 'built-in capacity of (plant) cells to proliferate' [39].

Cells in leaves of adult plants do not proliferate much *in situ*. To the contrary, when placed in culture conditions, these somatic cells proliferate, associate, and may recast as many tissues as a whole plant carries; this includes leaves like the ones from which the explanted single cell originated. It should be remembered that while plants are autotrophs, animal cells are auxotrophs, that is, the latter require complex nutritional components in their diet, whereas the former just require a source of carbon, nitrogen, salts, water and light.

In summary, from the available evidence pointing out that plant cells are auxotrophs that proliferate in defined medium [19], it follows that the default state in this kingdom, as in prokaryotes and unicellular eukaryotes, is *proliferation*.

## Conclusions

Experimental design, as well as the data emerging from it, are laden with theoretical meaning; that is, they are dependent on the premises chosen. In this chapter, we examined the historical context under which the control of cell proliferation has been studied. There is a consensus among biologists that the default state of unicellular organisms and metaphyta is *proliferation*. In contrast, the published record shows that the default state of cells in metazoa is assumed, but seldom stated, to be *quiescence*. However, no compelling reasons are given for this assumed evolutionary jump. During the advent of multicellularity both metaphyta and metazoa must have retained the ancestral default state, that is, *proliferation*. While cells isolated from metaphyta can be propagated in defined media, for the most part, cells from metazoa cannot be propagated, or even maintained alive for very long, in defined medium devoid of serum proteins. Hence, until the nutrition problem is solved, the default state of metazoan cells cannot be assessed experimentally.

The development of cell culture greatly facilitated the study of a good number of biological phenomena. The goal of achieving long-term propagation of cells from metazoa using a chemically defined serumless medium has been largely unsuccessful. This may be due to several reasons: first, the misreading of the data generated by Eagle and Piez which shifted away from the search for components in serum that played a nutritive role to that of *signals* having instead a stimulatory role in the proliferative event. In terms of research programs, this meant that axiomatically the default state became *quiescence* and the ultimate control was to be exerted positively (by growth factors); second, the failure of the above-mentioned approach resulted in a vacuum of knowledge, lasting up to the present, regarding the nutritional role of serum. These issues will not be solved until a serious effort is undertaken to understand the nutritive requirements of metazoan cells, and, more specifically, the nutritive role of serum.

From an evolutionary perspective, our own experimental data, and from inferences drawn from reinterpreting the data of others, we therefore propose that:

(i) *proliferation* is a constitutive property of cells, both in unicellular and multicellular organisms. Thus, their default state is *proliferation;*
(ii) cell proliferation can only be regulated by negative (inhibitory) signals;
(iii) the cell cycle is a series of algorithms (or reactions, or events) to make two cells out of one; it operates automatically and repetitively, as long as the nutrient supply is adequate and no inhibitors of cell proliferation are present or are recognized effectively. Once the cycle is started it cannot be naturally stopped until completion;
(iv) specific inhibitory signals move cells into quiescence ($G_0$). Control of cell proliferation operates as an on–off switch.

## References

1. **Jacob, F.** (1974) *The Logic of Life*. Pantheon Books, New York.

2. **Steinhaus, E.A. and Birkeland, J.M.** (1939) Studies on the life and death of bacteria. I. The senescent phase of aging cultures and the probable mechanisms involved. *J. Bacteriol.* **38**: 249–261.

3. **Soto, A.M. and Sonnenschein, C.** (1993) Regulation of cell proliferation: is

the ultimate control positive or negative? In: *New Frontiers in Cancer Causation* (ed. O.H. Iversen). Taylor & Francis, Washington, DC, pp. 109–123.

4. **Alberts, B., Bray, D., Lewis, J.G., Raff, M., Roberts, K. and Watson, J.D.** (1994) *Molecular Biology of the Cell*, 3rd Edn. Garland Publishing, New York, p. 891.

5. **Alberts, B., Bray, D., Lewis, J.G., Raff, M. and Roberts, K.** (1994) *Molecular Biology of the Cell*, 3rd Edn. Garland Publishing Inc., New York, pp. 880–881.

6. **Bonner, J.T.** (1988) *The Evolution of Complexity by Means of Natural Selection.* Princeton University Press, Princeton, NJ.

7. **Mayr, E.** (1982) *The Growth of Biological Thought: Diversity, Evolution, and Inheritance.* Belknap Press, Cambridge, MA, pp. 681–726.

8. **Varmus, H.E. and Weinberg, R.A.** (1992) *Genes and the Biology of Cancer.* Scientific American Library, New York, p. 123.

9. **Gould, S.J.** (1977) *Ever Since Darwin: Reflections on Natural History.* Norton Publishing, New York, pp. 91–96.

10. **Williams, G.T., Smith, C.A., Spooncer, E., Dexter, T.M. and Taylor, D.R.** (1990) Haemopoietic colony stimulating factor promotes cell survival by suppressing apoptosis. *Nature* **343**: 76–79.

11. **Wyllie, A.H., Kerr, J.F. and Currie, A.R.** (1980) Cell death: the significance of apoptosis. *Internat. Rev. Cytol.* **68**: 251–306.

12. **Leist, M. and Nicotera, P.** (1997) The shape of cell death. *Biochem. Biophys. Res. Commun.* **236**: 1–9.

13. **McKenna, S.L. and Cotter, T.G.** (1997) Functional aspects of apoptosis in hematopoiesis and consequences of failure. *Adv. Cancer Res.* **71**: 121–164.

14. **Gilbert, S.** (1997) *Developmental Biology*, 5th Edn. Sinauer Associates, Sunderland, MA, pp. 591–634.

15. **Ronnov-Jessen, L., Peterson, O.W. and Bissell, M.J.** (1996) Cellular changes involved in conversion of normal to malignant breast: importance of the stromal reaction. *Physiol. Rev.* **76**: 69–125.

16. **Bryant, P.J. and Schmidt, O.** (1990) The genetic control of cell proliferation in Drosophila imaginal discs. *J. Cell Sci.* (Supplement) **13**: 169–189.

17. **Bryant, P.J.** (1997) Junction genetics. *Developmental Genetics* **20**: 75–90.

18. **White, P.R.** (1934) Potentially unlimited growth of excised tomato root tips in a liquid medium. *Plant Physiol.* **9**: 585.

19. **Willmer, E.N.** (1965) Introduction. In: *Cells and Tissues in Culture: Methods, Biology and Physiology* (ed. E.N. Willmer). Academic Press, London, pp. 1–17.

20. **Willmer, E.N.** (1965) *Cells and Tissues in Culture: Methods, Biology, and Physiology.* Academic Press, Cambridge, MA.

21. **Moscona, A.** (1952) Cell suspensions from organ rudiments of chick embryo. *Exp. Cell Res.* **3**: 535–539.

22. **Rous, P. and Jones, F.S.** (1916) A method of obtaining suspensions of living cells from the fixed tissues and from the plating out of individual cells. *J. Exp. Med.* **23**: 549–555.

23. **Baserga, R.** (1976) *Multiplication and Division of Mammalian Cells.* M. Dekker, New York.

24. **Waymouth, C.** (1965) Construction and use of synthetic media. In: *Cells and Tissues in Culture: Methods, Biology, and Physiology* (ed. E.N. Willmer). Academic Press, Cambridge, MA, pp. 99–121.

25. **Rose, W.C.** (1938) The nutritive significance of the amino acids. *Physiol. Rev.* **18**: 109.

26. **Eagle, H. and Piez, K.A.** (1960) The utilization of proteins by cultured human cells. *J. Biol. Chem.* **235**: 1095–1097.

27. **Kent, H.N. and Gey, G.O.** (1957) Changes in serum proteins during growth of malignant cells in vitro. *Proc. Soc. Exp. Biol. Med.* **94**: 205–208.

28. **Levi-Montalcini, R.** (1986) The nerve growth factor: thirty-five years later. *The Nobel Lectures* 279–299.

29. **Cohen, S.** (1986) Epidermal growth factor. *The Nobel Lectures* 263–275.

30. **Cohen, S. and Elliot, G.A.** (1962) The stimulation of epidermal keratinization by a protein isolated from the submaxillary gland of the mouse. *J. Invest. Dermat.* **49**: 1–5.

31. **Barnes, D. and Sato, G.H.** (1980) Serum-free cell culture: an unifying approach. *Cell* **22**: 649–655.

32. **Taylor, W.G.** (1974) 'Feeding the baby': Serum and other supplements to chemically defined medium. *J. Natl Cancer Inst.* **53**: 1449–1457.

33. **Alberts, B., Bray, D., Lewis, J.G., Raff, M., Roberts, K. and Watson, J.D.** (1994) *Molecular Biology of the Cell*, 3rd Edn. Garland Publishing, New York, p. 159.

34. **Hayflick, L.** (1992) Aging, longevity and immortality. *Exp. Gerontol.* **27**: 363–368.

35. **Alberts, B., Bray, D., Lewis, J.G., Raff, M., Roberts, K. and Watson, J.D.** (1994) *Molecular Biology of the Cell*, 3rd Edn. Garland Publishing, New York, pp. 903–904.

36. **Tauber, J.P., Cheng, J., Massoglia, S. and Gospodarowicz, D.** (1981) High density lipoproteins and the growth of vascular endothelial cells in serum-free medium. *In Vitro* **17**: 519–530.

37. **Moore, G.E. and Minowada, J.** (1969) Human hemapoietic cell lines: a progress report. In: *Hemic Cells in Vitro* (ed. P. Franes). The Williams & Wilkins Co., Baltimore, pp. 100–114.

38. **Rubin, H.** (1997) Cell aging in vivo and in vitro. *Mechanisms of Ageing & Development* **98**: 1–35.

39. **Steward, F.C., Kent, A.E. and Mapes, M.O.** (1966) The culture of free plant cells and its significance for embryology and morphogenesis. *Current Topics in Developmental Biology* **1**: 113–154.

# Chapter 3

# The Experimental Evaluation of Cell Proliferation

"One must show the greatest respect toward anything that increases exponentially no matter how small."

*Garret Harding (1968) Exploring New Ethics for Survival: The Voyage of the Spaceship Beagle. Viking Press, New York, p. 45*

". . . I have serious reasons to believe that the little prince's planet of origin was the asteroid known as B-612. This asteroid has only been observed once through a telescope by a Turkish astronomer in 1909.
At the time, he organized a great demonstration of his discovery at an International Astronomical Congress. But, because of his Turkish attire, nobody believed him. Grown-ups are like that.

Fortunately for the reputation of Asteroid B-612, however, a Turkish dictator imposed European costume upon his subjects under pain of death. So the astronomer repeated his demonstration in 1920, dressed in an elegant suit. And this time, everybody was convinced . . ."

*Antoine de Saint-Exupery (1995) The Little Prince. Wordsworth Classics, Hertfordshire, UK, pp. 20–21*

## Introduction

The study of the control of cell proliferation, like any other area of experimental science, depends on at least two types of prerequisites: first, the development of hypotheses to be tested, and second, the development of appropriate tools and methods with which to test them.

Once the cell theory was accepted, it became evident that the growth of multicellular organisms depended on the generation of new cells and extracellular components. The discovery of mitosis provided the means of identifying cells undergoing division. However, mitosis represents a relatively short process in the cycle, lasting about 1 hour or less (roughly, 5% or less of the total length of the cycle). Hence, counting how many cells were undergoing mitosis in a given tissue at the moment that the tissue was frozen in time by removing it from the animal and fixing it, provided a means of assessing and comparing the proliferative activity of this given tissue in various experimental conditions. The study of how organs increased in size, and how the adult size of an organ was maintained through the processes of cell proliferation and cell death was laboriously worked out in the first 70 years of this century. Technical innovations such as radiolabeled DNA precursors and flow cytometry, and the discovery of programmed cell death (apoptosis) allowed for more accurate descriptions of these processes. Computers are now being used in morphometrics to produce more precise descriptions of how organs undergo development and growth. However, the study of why cells proliferate cannot be explored at the organismal level of complexity. Thus, it is not surprising that most of the work in this area has been done using cells *in culture*.

Bacteria and unicellular organisms proliferate exponentially as long as they are maintained in the presence of excess nutrients. Similarly, metazoan cells proliferate exponentially in culture conditions. This chapter will evaluate the soundness of the methodology used to measure cell proliferation.

### *Methods to evaluate cell proliferation in tissues of metazoa*

The study of cell proliferation may be parceled into three periods: first, during organ formation, second, during growth to adult size, and third, once adult size is attained. Studying these phenomena *in vivo* is a difficult task, because of the many tissues present in an organ, the different cell types present in a tissue, and the complexity of the three-dimensional architecture of the organ or tissue studied which contains both cells and extracellular substances (extracellular matrix). Quantitative studies at the tissue level can now be done using a variety of complex techniques, involving computer-assisted mathematical modeling. This methodology is still evolving [1]. We will concentrate on the simple, not quite quantitative studies that are done routinely. Equipped with a common light microscope and a few tools, researchers may ask the question: does agent $x$ affect the proliferation rate of cells in tissue $y$? Then, the effective dose of agent $x$ and the time-course of its action should be found. Only two phases of the cycle can be revealed at the light microscope, the S phase when DNA synthesis takes place, and the M phase, when mitotic figures can be seen. To verify whether or not the agent being studied had affected the proliferation of a given tissue, one follows the time-course of the incorporation of the precursor into the DNA, measuring the percentage of labeled cells, and of cells in mitosis. Questions about whether or not there is a population of quiescent cells can only be studied by administering a DNA precursor for a prolonged period of time (the length of several cell cycles), and then counting the percentage of cells that remain unlabeled.

### *Methods used to evaluate proliferation of cells in culture*

Proliferation is a multistep process from which one cell generates two cells; consequently, measuring cell proliferation entails measuring the accrual of cells as a function of time; this is done by simply counting cells [2]. For over two decades now, counting cells has become a routine operation easily done by means of electronic particle counters. This is the only direct way of ascertaining the accumulation of new cells. Alternatively, indirect methods are used whereby cells are stained *in situ* with dyes that bind to proteins or to DNA; next, the dye is solubilized and its absorbance measured. The amount of dye should be proportional to the amount of cells in the dish. However, there may be situations in which cells may enlarge, hence, the amount of protein per cell may be influenced independently of the increase in cell number. Thus, prior to the use of a dye to stain proteins, experimental verification of the accuracy of the procedure should take place. Finally, the amount of dye is converted to a cell number [3]. It remains factual that the most accurate method is that of counting particles, that is, cells or their nuclei.

### *Proliferation rate experiments*

In the presence of abundant nutrients, metazoan cells in culture, like bacteria, proliferate in a pattern comprised of three phases (*Figure 3.1a*). First, there is a lag period, immediately after cells are subcultured. During this period, cells do not proliferate, presumably because they have to overcome the shock caused by the procedures used to start a new passage (forceful pipetting to disperse cell clumps, trypsin treatment to detach them from the walls of the dish, etc.). Second, a log phase occurs where cells proliferate following a geometrical progression of factor 2; the duration of this phase depends on the availability of nutrients and the inoculum size (cultures started from small inocula last longer within the exponential phase than do those from larger inocula). Finally, there is a stationary phase, where no further increase in cell number occurs. During this period, as nutrients become exhausted and/or cells become crowded, the proliferation rate slows down. Most importantly, proliferation rates should be measured within the log phase; it is during this period that cells will show whether or not the experimental protocol being tested alters their rate of proliferation.

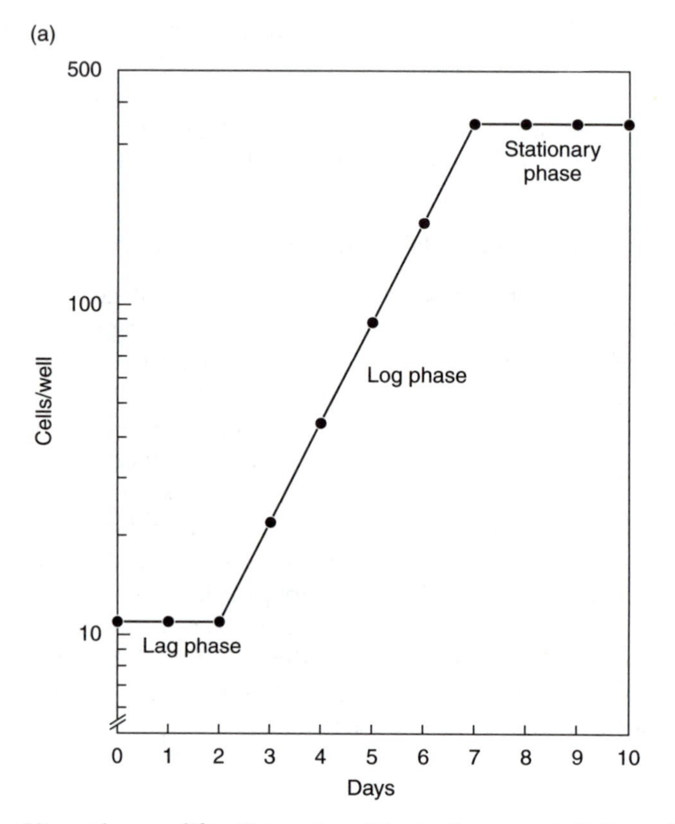


**Figure 3.1a.** Measuring proliferation rates. Schematic representation of the three stages of a proliferation curve. Time 0 indicates the time of seeding.

Cell proliferation follows a simple equation for exponential behavior: $C_t/C_0 = e^{(\alpha-\rho)t}$, where $C_t$ is cell number at time = $t$, $C_0$ is cell number at $t$ = 0, $\alpha$ is the instantaneous proliferation rate constant and $\rho$ is the instantaneous death rate constant. Cell proliferation is best evaluated by measuring the doubling time ($t_D$) of a cell population. $t_D$ is the time interval in which an exponentially growing culture doubles its cell number; there is only one case in which doubling time is equivalent to generation time, or mean duration of cell cycle, namely, the fraction of proliferating cells should be 100% and cell death rate should be 0. Determining $t_D$ requires measuring cell yields at several time intervals during the exponential proliferation phase. Hence, data supporting claims about a growth factor or a proliferation inhibitor should show a significant difference on the slope of the proliferation curve, and thus, on the $t_D$ when compared to control conditions (*Figure 3.1b*).

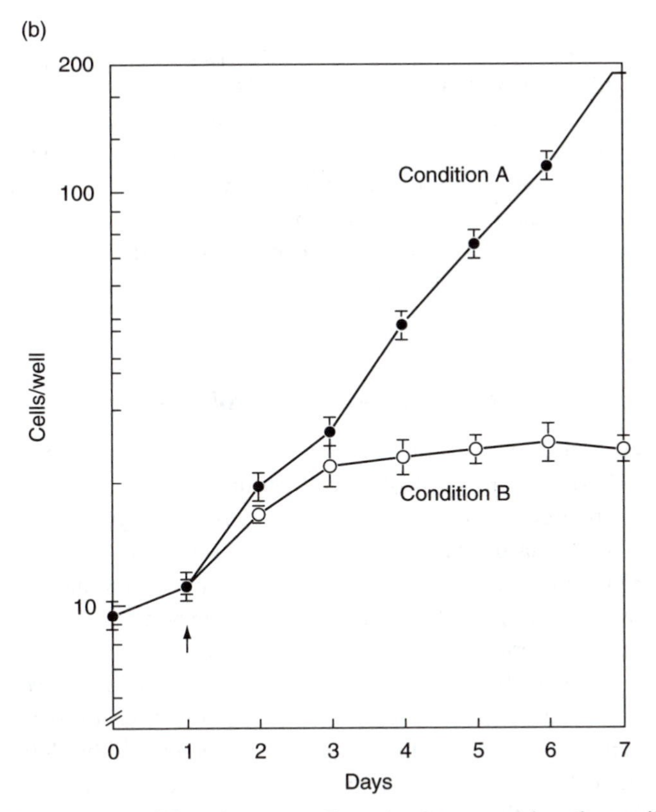


**Figure 3.1b.** Measuring proliferation rates. Experiment comparing the proliferation of similar inocula of a cell line in two different experimental conditions. Time 0 indicates the time of seeding. The cells were exposed to the test media one day after seeding (this is indicated by the arrow). Condition A: chemically defined medium. Condition B: chemically defined medium plus 10% serum made free of sex steroids by adsorption with dextran-coated charcoal.

### *Cell yield experiments*

A less cumbersome alternative to measuring proliferation rates is comparing the cell yield achieved by similar cell inocula harvested simultaneously at a single point during the exponential phase of proliferation. Obviously, the experimentalist has to first find the right conditions (inoculum size, length of the exponential phase) before deciding when cells ought to be harvested. Proliferation rate and cell yield experiments only provide data on the behavior of the cell population (whether or not there is a net increase of the cell number during the period studied, and of the rate of accumulation of new cells). However, if a population does not increase in number during a time interval, this method cannot discriminate between cells that are quiescent, and a pattern whereby each cell generates one cell that dies and another that survives and proliferates again with the same pattern. When using 'monolayer' cultures, most cell types attach well to the bottom surface of the dish, while dead cells detach from the dish surface, become 'floaters', and are washed out and, hence, not counted. For cell types that proliferate in suspension, the population counted has both viable (live) and dead cells. Hence, questions about the percentage of dead cells, cells that traverse the cell cycle and those that do not (quiescent cells) are answered by the methodology outlined below.

### *Cell cycle analysis as a tool to address cell proliferation*

The hypotheses chosen to study a phenomenon influence the experimental design. Hypotheses on the control of cell proliferation in metazoan cells differ in two aspects, namely, on whether or not there is a state of true quiescence (the $G_0$ state), and on the nature of the default state. If the default state is construed to be *quiescence* by nutrient (usually serum) deprivation, metazoan cells can be made quiescent *in culture*. This is akin to cell starvation. Obviously, the full cast of characters present in a plastic dish is not equivalent to the *in-situ* situation in the animal. Cells in the organism are bathed by a milieu that contains serum proteins. Instead, serum-starved cells *in culture* have been used to characterize putative growth factors. The putative growth factor should be defined by assessing how many cells have traversed the cell cycle in its presence, and by measuring the time-course of this event.

To the contrary, if the default state is construed to be *proliferation*, cells should proliferate maximally in medium containing all the necessary nutrients. When exposed to the putative inhibitor, they should cease to proliferate. Then, the time-course of entry into quiescence is followed, as well as the percentage of cells that undergo quiescence. The competing positive and negative hypotheses acknowledge that control of cell proliferation entails the decision of entering the cycle or entering quiescence, respectively. Most importantly, this decision is taken at a hierarchical level different (cellular) from that at which control of cell cycle traverse takes place, that is, the subcellular level. Therefore, in this context, the purpose of studying cell cycle kinetics is to establish whether or not there is a population of quiescent cells, and to quantitate the movement of cells from one cycle compartment to another (entering and leaving the cycle).

The notion that there is no true quiescence was conceived by cell kineticists that studied the renewal of epithelial tissues *in situ*. For example, the basal layer of the epidermis is continuously generating new cells. If an animal is injected frequently with a labeled DNA precursor for several days, all cells in the basal layer become labeled. Hence, all these cells appear to be in the pool of proliferating cells. In contrast, other organs, such as the adult liver, have a large population of quiescent cells. Hence, these data led most researchers to adopt the notion that quiescence is a physiological state.

Researchers that assume that the default state is *proliferation*, but reject the existence of a truly quiescent state ($G_0$), and therefore of a switch, assume that the control of cell proliferation is exerted by controlling the velocity of cell cycle traverse. They look for changes on the duration of the phases of the cycle, especially of $G_1$ [4].

### *Measuring cell death*

Cell death is a phenomenon conceptually independent from that of the control of cell proliferation. Cell death plays a role in morphogenesis, and on the control of organ cell numbers [5]. *In culture*, cell death may occur by damage due to harsh environmental conditions, such as nutrient depletion or mechanical disruption (necrosis), or due to a physiological phenomenon initiated within the cell (programmed cell death or apoptosis).

Death of cells *in culture* resulting from damage due to manipulation (thawing frozen cells, pipetting, etc.) may be assessed by the incorporation of vital dyes, like trypan blue; while living cells take up and release this dye, dead cells do not release it and hence become colored. Death by apoptosis may be evaluated by the presence of morphological or biochemical markers. Biochemical markers of apoptosis, such as incorporation of nucleotides at the 3′ end of the fragmented DNA by action of the enzyme terminal deoxynucleotide transferase, can also be detected by microscopy. Cells that are undergoing apoptosis incorporate a labeled nucleotide that can be visualized in cytochemical assays. Hence, the percentage of cells undergoing apoptosis may be measured. In some cell types, apoptosis can also be detected by flow cytometry.

## Fitting data into working hypotheses

We have chosen the premise stating that the default state in all living cells is *proliferation*. After a tortuous itinerary, we identified a serum component that specifically inhibits the proliferation of a set of well-defined target cells. First, we measured the proliferation rate of these target cells in the absence of the suspected inhibitor. Cells were harvested daily and the length of the exponential phase was determined. Next, diverse concentrations of suspected inhibitor were tested for their ability to alter the doubling times of the target cells. Then, we assessed whether the inhibitory effect of accrual of new cells was due to quiescence or cell death. We concluded that our agent inhibited cell proliferation. Finally, we studied the effect of removal of the inhibitor, first by proliferation curves, and then, by cell cycle analysis. We observed that inhibition was reversible. In this clear-cut case, we did not have to deal with cell death, since the changes in the doubling time could be attributed to most cells entering a quiescent state (see Chapter 5).

Now, we will briefly switch to a totally different case, that is, that of the proliferation of erythropoietic cells, the precursors of erythrocytes. Bleeding, as well as exposure to low levels of oxygen (such as those existing at high altitudes), results in an increased rate of red blood cell production. It is also known that a hormone secreted by the kidneys, erythropoietin (EPO), is the mediator of this effect. When the effect of EPO was tested by means of proliferation curves, it was observed that the cell number increased only in the presence of this hormone. Thus, it was assumed at first that EPO was a growth factor. However, later on, it was found that precursor cells in the absence of EPO were not just quiescent in $G_0/G_1$ phase, but they actually were dying. This was demonstrated by measuring DNA degradation [6, 7]. Hence, EPO does not move cells from quiescence to proliferation, but rather it inhibits their death, thus allowing the cells to express their constitutive ability to proliferate.

## The use and misuse of tritiated thymidine incorporation data

The use of the DNA precursor tritiated thymidine to assess cell proliferation is a common practice in *animal* experiments. This stems from the fact that in order to proliferate, cells have to traverse the cell cycle, and thus, synthesize DNA. The incorporation of tritiated thymidine may be assessed histologically by autoradiography; silver grains in the nuclei of cells identify those that have incorporated the precursor into DNA. Proliferation is a discrete cellular function; thus, being able to document which cells in a population undergo DNA synthesis is, in principle, a sound strategy. In practice, the most stringent parameter to assess the S phase is to measure the DNA content of the cells, since cells in S phase have an intermediate DNA content between 2N and 4N.

More frequently, incorporation of radioactive precursor into the DNA of cells *in culture* is assessed by exposing them to the labeled DNA precursor for 1 or 2 hours, lysing them, and measuring the radioactivity incorporated into DNA. However, tritiated thymidine incorporation into DNA is the result of at least three discrete events: namely, the ability to enter the cell, to be metabolized into the nucleotide, and, finally, to be incorporated into DNA. An increase in thymidine incorporation into DNA may occur when the rate of any one of these three events is increased, thus leading to spurious conclusions [8, 9]. Moreover, this method cannot assess whether incorporation of precursor into DNA represents DNA duplication or DNA repair. Hence, the use of this parameter to assess cell proliferation is inadequate, unless the occurrence of effects other than increased DNA synthesis, due to DNA replication, are ruled out. First, it has to be determined whether or not the agent that is used to induce, or to inhibit, cell proliferation affects the uptake or metabolism of thymidine, independently of its suspected effect on cell proliferation. Then, it has to be determined whether DNA synthesis represents genuine replication. At best, when precursor incorporation into DNA is due to the actual proliferative activity of cells, the radioactivity incorporated cannot be translated quantitatively into cell proliferation parameters, such as doubling time. This is because cell proliferation is a discrete function (one cell generates two cells, rather than one and a half), and incorporation of a precursor into DNA is a continuous function during the time that DNA is synthesized (a given amount of nucleotides are incorporated per unit time, and per unit of DNA, until all the DNA is copied). Hence, thymidine incorporation cannot be converted into parameters such as doubling time or percentage of cells in S phase. For the most part, reports using thymidine incorporation into DNA as a method for assessing cell proliferation fail to clarify whether or not the spurious effects discussed above have been ruled out experimentally.

## Conclusions

Cell proliferation is tightly regulated in multicellular organisms during development and throughout adult life. Studies using animal models reveal that the processes of histogenesis, organogenesis and maintenance of cell numbers during adulthood involve the interaction of diverse cell types and signals. However, it is practically impossible to assess causality in a highly interactive model. For this reason, proliferation control is mainly studied *in culture*, where in most cases, a single cell type is present, and a single extracellular variable is tested.

Cell proliferation is a discrete function. The only parameter that allows comparisons of proliferative activity is the doubling time of a population. If one accepts that the default state is *quiescence*, the regulatory agent (growth factor) must significantly decrease the doubling time. If, instead, one accepts *proliferation* as the default state, the controlling agent (inhibitor) must significantly increase the doubling time. Once this is documented, one may ask 'How has this happened?' Researchers would investigate whether this was the result of cells entering or leaving the proliferative pool, and/or of cell death. The validation of these *in-culture* findings will require the development of a research program aimed at verifying, in the animal, whether or not the role of the putative proliferative regulator is consistent with the data obtained in culture conditions.

## References

1. **Howard, C.V. and Reed, M.G.** (1998) *Unbiased Stereology.* BIOS Scientific Publishers, Oxford.

2. **Sonnenschein, C. and Soto, A.M.** (1980) But . . . are estrogens per se growth-promoting hormones? *J. Natl Cancer Inst.* **64**: 211–215.

3. **Baserga, R.** (1995) Measuring parameters of growth. In: *Cell Growth and Apoptosis: A Practical Approach* (ed. G.P. Studzinski). IRL Press, Oxford, pp. 1–19.

4. **Iversen, O.H.** (1961) The regulation of cell numbers in epidermis: a cybernetic point of view. *Acta Pathol.* **148**: 91–96.

5. **Gilbert, S.** (1997) *Developmental Biology,* 5th Edn. Sinauer Associates, Sunderland, MA, pp. 724–727.

6. **Williams, G.T., Smith, C.A., Spooncer, E., Dexter, T.M. and Taylor, D.R.** (1990) Haemopoietic colony stimulating factor promotes cell survival by suppressing apoptosis. *Nature* **343**: 76–79.

7. **Koury, M.J. and Bondurant, M.C.** (1988) Maintenance by erythropoietin of viability and maturation of murine erythroid precursor cells. *J. Cell Physiol.* **137**: 65–74.

8. **Drach, J.C., Thomas, M.A., Barnett, J.W., Smith, S.H. and Shipman, C. Jr** (1981) Tritiated thymidine incorporation does not measure DNA synthesis in ribavirin-treated human cells. *Science* **212**: 549–551.

9. **Wolff, S. and Bodycote, J.** (1986) Metabolic breakdown of [$^{3}$H] thymidine and the inability to measure human lymphocyte proliferation by incorporation of radioactivity. *Proc. Natl Acad. Sci. USA* **83**: 4749–4753.

## Chapter 4

# Hypotheses for the Control of Cell Proliferation

"False facts are highly injurious to the progress of science, for they often endure long; but false views, if supported by some evidence, do little harm, for every one takes a salutary pleasure in proving their falseness."

*Charles Darwin, quoted in: Steven Jay Gould (July 1996) Natural History* ***105****: 16*

"What gets us into trouble is not what we don't know. It's what we know for sure that just ain't so."

*Jogi Berra*

## Historical perspective

The study of the control of cell proliferation in metazoa stems from many different scientific traditions. We will explore only three of them. Pathologists were concerned with the regeneration of residual parts of an organ after resection of a portion of it. Almost concurrently, endocrinologists were investigating the trophic role of hormones. Finally, cell biologists were developing methods to propagate and maintain cells and tissues in culture conditions. They all brought into the field of the control of cell proliferation their respective insights and shortcomings.

### *The pathologist's tradition*

In rodents, the removal of two-thirds of the liver is followed by the rapid restoration of the organ mass within the next 48–72 hours, with a complete remodeling in about a week. A number of researchers investigated which signals mediate this process. They all agreed that the signal was coming from the liver cells themselves: however, they differed significantly in their views on its mechanism. Was cell proliferation in the liver due to a positive signal produced by the functional remaining parenchyma or, was it due to a sudden drop in the blood concentration of a negative signal produced by the remaining liver cells? The positive option assumed that *quiescence* was the default state of these cells, while the negative option was based on the premise that *proliferation* was the default state.

The positive option is the more complex one, for it requires, first, the induction of a proliferative signal by the cells in the stump, and second, a negative signal for proliferation to cease when the normal mass is reached. In addition, this positive option begs another question: what hypothetical signal would indicate to cells in the remaining organ that there was a loss of parenchyma and, as a result, that they should start producing liver-specific growth factors? (see *Figure 4.1*).

The negative hypothesis proposes, instead, that the concentration of a putative inhibitor of liver cell proliferation would be maintained constant at effective plasma levels in intact animals. Immediately after the partial hepatectomy, the inhibitor level would decrease, triggering the proliferation of the remaining cells. Finally, as the cell numbers increase to reach those present in intact controls, the plasma levels of liver-specific inhibitor would increase, thus shutting off the proliferation of liver cells (*Figure 4.1*).

Complex experiments connecting the blood supply of partially hepatectomized and intact animals (parabiosis) resulted in data consistent with both hypotheses.

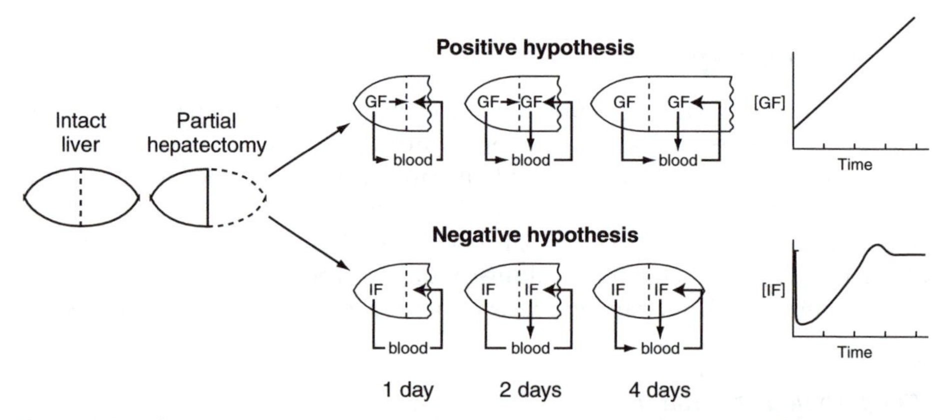


**Figure 4.1.** Schematic representation of liver regeneration after partial hepatectomy according to competing hypotheses for the control of cell proliferation. The positive control hypothesis does not account for a predictable end of the proliferative period once the liver mass is restored to its pre-hepatectomy cell number status. The plasma concentration of an inhibitory factor (negative control hypothesis) is proportional to the liver cell number and parsimoniously accounts for the initiation and cessation of cell proliferation [2].

This controversy has not been resolved to this day; this is primarily due to the absence of a bioassay which can reliably assess the purification of putative hepatocyte growth factors, or inhibitory factors [1]. The difficulty in generating data compatible with either explanation encouraged researchers to offer theoretical resolutions for this puzzle [2].

### *Paul Weiss' theory of growth/proliferation control*

It is worth spending some time on the genesis of Weiss' template–antitemplate theory because it illustrates how beliefs, postulates, and provisional tenets influence the thoughts of scientists. Weiss regarded homologous compensation, that is, the growth of a tissue after partial extirpation, as a process guided by the plasma levels of an inhibitor produced by the same cell type. The way he dissected the problem illustrates the temptation of scientists to accommodate 'all the facts' across different hierarchical levels, even when contradictory. It is generally assumed that, if nothing is left out, the resolution of the problem may be closer to the (elusive) truth. An alternative, and perhaps riskier approach, may be more sound in the long run; that is, hypotheses and published data should, instead, be ranked according to their plausibility, which is admittedly a subjective undertaking. Despite this subjectivity, it is as worthy today as it was when Kepler decided to chart the orbit of planets by using only the data obtained by his teacher, Tycho Brahe.

Weiss assumed that the unit of organ mass reproduction was the protoplasm, rather than the cell [3] (see Chapter 1). Since hypertrophy was claimed to precede cell proliferation during liver regeneration, he considered both processes as part of the same compensatory mechanism. Therefore, he was concerned with organ growth rather than with control of cell proliferation. He also accepted as valid claims that tissue extracts would induce the growth of their homologous tissues when injected into embryos. He then attempted to reconcile these notions with previous data showing that partial hepatectomy triggers a specific proliferative response. In order to integrate these contradictory sets of data, he proposed that, first, the growth rate of a tissue is proportional to the presence in their cells of growth catalysts named 'templates', and second, the cells produce tonically tissue-specific growth inhibitors, which he called 'anti-templates'. These inhibitors were supposed to be secreted into the bloodstream and hence, reach all organs, including the one producing them. Target organs would recognize and internalize the 'anti-template'. The intracellular interaction between 'templates' and 'anti-templates' would then stop growth[1].

---

[1] For readers who may shrug off the notion of template/anti-template, we make reference in Part II, Chapter 8 to a more modern version of this idea when we analyze the plausibility of the oncogene/anti-oncogene (suppressor gene) couple.

In summary, the pathologists acknowledged that negative control best fits the data on control of cell proliferation during organ regeneration. Weiss, a developmental biologist, focused his analysis on organismic and organ-based hierarchical levels (growth control), and not on the cellular level (cell proliferation control). Had he focused on cell proliferation rather than on organ growth, he probably would have concluded that the default state of all cells was *proliferation*.

*The 'fin de siecle' version of hepatic regeneration*

The prevalent hypothesis on hepatic regeneration proposes that growth factors are the main signals in the process of repopulating the partially resected liver [1]. So far, those who explore this hypothesis have ignored the epistemologic objections outlined above. Notwithstanding, hepatocyte growth factor (HGF) and transforming growth factor alpha (TGF-α) have been singled out as the leading representatives of this group. The proliferative role assigned to HGF is challenged by results using mice homozygous for a null mutation of the HGF gene (HGF knockouts). This lethal mutation produced discordant phenotypes regarding its effects on liver development. One publication reports normal liver architecture at 13.5 days of intrauterine life [4]. The other publication reports reduced liver size measured at 12.5 days of intrauterine life, attributed to cell death and dissociation of parenchymal cells, rather than to impaired proliferation [5]. In fact, these hepatocytes proliferated uneventfully in culture medium lacking HGF. In addition, researchers candidly admit the temporal inconsistencies and lacks of fit that become evident when linking HGF and other growth factors with the cause of hepatocyte proliferation [6, 7]. When dealing with a proliferative event based on the implicit premise that the default state of cells in multicellular organisms is *quiescence*, researchers face the unavoidable questions, 'What may shut down the positive signals?', or specifically, 'What makes liver cells stop proliferating when they eventually reach the right number?' The alternative proposed is the introduction of *ad hoc* inhibitors of cell proliferation to close this loophole. TGF-β1 has been postulated as the main candidate for this role. However, this possibility is challenged by reports claiming it to be a strong 'mitogen' for mouse and rat hepatocytes [1]. The mechanisms underlying hepatic regeneration are still unknown.

### *The endocrinologist's tradition*

The ablation of endocrine glands also generated interesting insights into the control of cell proliferation. From dwarfism, after the removal of the hypophysis, to the atrophy of accessory sex organs (uterus, prostate, and others) after gonadal ablation, these models have been used extensively. These phenomena led to the discovery of hormones, chemical signals that exert their trophic activity in distant organs. In contrast to the 'tissue regeneration' perspective that postulated inhibitory signals, endocrinologists discovered trophic signals that seemed to act in a stimulatory fashion. However, they also discovered the mechanism of negative feedback control; this concept explained why blood glucose levels could be kept within a relatively narrow range. Negative feedback also explained why levels of a trophic hormone secreted by the hypophysis (e.g. luteinizing hormone) were kept in check by the target cell product (in this case, testosterone) secreted by the testis. However, some vexing problems remained that could not be explained in the context of trophic hormones and negative feedback. For example, once the prostate reaches adult size, cell proliferation stops despite high plasma levels of the trophic hormone, testosterone, the proximate cause of cell proliferation in this organ.

Both the 'negative' and 'positive' traditions reached a point where data gathered in animal models were unsuitable to answer questions about the ultimate cause of the cellular proliferative effects apparently triggered by these hormones. Beginning in the early 1970s, advances in cell culture techniques facilitated the study of hormonal effects on gene expression as well as on cell proliferation (see Chapter 5).

### *The tissue/cell culture tradition*

The use of *in-culture* models grew from the perceived inadequacy of whole animal models to generate answers to questions on causality. The availability of purified hormones and of 'established' cell lines encouraged researchers to explore the mechanisms of hormone action. From our analysis, we conclude that the overriding premise under which these studies were conducted is that hormones are direct stimulators of gene expression *and* cell proliferation. As far as cell proliferation is concerned, this conclusion further implies that the default state of cells *in-culture* is *quiescence.* Thus, tacitly, the assumption has been that cells had to be stimulated to enter the cycle [8]. Much has been written about how data stemming from *in-culture* experiments should be integrated into the complexity of biology at large [9]. A renewed debate over this subject is now warranted.

## Positive hypotheses

### *The positive hypothesis for the control of cell proliferation*

On what premises does the positive hypothesis for the control of cell proliferation rest? The premises are: first, the default state of metazoan cells is *quiescence;* second, the state of quiescence of cells in multicellular organisms is equivalent to the state of quiescence of cells *in culture* subjected to serum starvation; third, in culture conditions, the 'defined' medium contains all the nutrients needed for survival; and finally, serum provides both metabolic regulators (hormones, like insulin that facilitates nutrient uptake) and *signals* to induce cell proliferation (growth factors) [10]. The notion of *quiescence* in metazoa has acquired an axiomatic quality and, thus, it is seldom discussed. From this rationale, the need for stimulators of cell proliferation becomes inevitable.

### *Growth factors*

In 1976, Gospodarowicz and Moran listed a number of requirements that would validate the existence of any novel growth factor [11]; this represents a rare instance in which the proponents of the positive hypothesis defined what a growth factor ought to be. The requirements were: first, growth factors should induce several cycles of cell proliferation both in sparsely seeded and in confluent cell cultures; second, they should support clonal growth and indefinite propagation of their target cells; and, third, the specificity of their proliferative action should be demonstrated in the animal. In other words, the evidence collected in culture conditions should be matched by a comparable physiological proliferative role. Instead, growth factors were defined *operationally* as polypeptides whose role was to trigger the initiation of the cycle in cells under nutritionally restrictive conditions. Baserga nodded cautiously, '. . . this is not to say that reproduction *in vivo* is regulated by the same factors, but cell cultures are where we must start' [12].

Have 'growth factors' met the requirements listed by Gospodarowicz and Moran? Given the voluminous literature on growth factors, we cannot exhaustively review data on all of them, nor can we realistically refute all arguments regarding their merits. However, a few general comments will be made. For instance, data portraying a proliferative effect by growth factors are supported, for the most part, by the increase of tritiated thymidine incorporation into the DNA of cells exposed to these putative positive signals when compared with 'non-stimulated' controls. Generally, when proliferation curves are shown they are not exponential. We addressed our concerns over this evidence in Chapter 3.

The advent of recombinant DNA technology made possible the use of species-specific recombinant growth factors, and the generation of mice carrying null mutations (knockouts) of putative growth factors and their specific receptors. In the words of Durum and Muegge, the introduction of this technology provides the desired 'acid test for the function of a gene' and consequently, validates claims emanating from data gathered *in culture* [13]. The data collected so far have failed to clarify whether or not growth factors have a role on the control of cell proliferation. A clarification is in order: the data stemming from work on developmental genetics suggest that these polypeptides may indeed play roles as morphogens or differentiation factors [14]. Also, signal transduction pathways may be activated as a result of an interaction between these putative growth factors with their specific plasma membrane receptors. Here, we are just challenging the notion that these molecules are instructive signals for cells to enter the cycle.

### *The epidermal growth factor (EGF) story*

Work on EGF evolved from work on nerve growth factor (NGF). Levi-Montalcini described NGF as a substance that, when added to sympathetic nerve ganglion cells, increased the number of their cytoplasmic processes (neurites). Recent evidence indicates that NGF is a 'survival' factor by virtue of its ability to prevent apoptosis in neurons and lymphocytes [15, 16].

Cohen, who collaborated with Levi-Montalcini on NGF research, reported in the early 1960s that when injected into newborn animals, extracts of male mouse submaxillary glands induced the precocious opening of their eyelids and early tooth eruption. Curiously, extracts of female glands appeared ineffective. Eyelid opening was used as the assay for the eventual purification of EGF. Interestingly, the 'stimulated' pups gained less weight than the controls injected with vehicle. It was first assumed that the thickening of the epidermis observed in EGF-injected pups was due to a proliferative effect. However, cell kinetics studies in organ culture revealed that '. . . an increase in the number of epidermal layers is noted in the experimental cultures but without any appreciable increase in the number of ($^3$H thymidine) labeled basal cells . . .' [17]. This evidence compelled Cohen to conclude that '. . . the factor initially stimulates the migration and piling up of the epidermal cells, and *secondarily* increases the number of mitosis in the basal layers' (our italics). Green's group investigated the effects of EGF on the proliferation of keratinocytes [18]. Their data concurred with those of Cohen. EGF behaved as a 'migrating factor' of keratinocytes *in culture*. When EGF was added to the culture medium, spreading allowed for more cells to be present on the periphery of a colony. Only the cells on the periphery of the colonies proliferated. In summary, the effect of EGF on the proliferation of these cells appeared to be **indirect**.

The lack of a proliferative, physiological role for EGF in animals generated a wave of *ad hoc* explanations. For instance, it was suggested that EGF operated at a local level (paracrine and autocrine mechanisms). These mechanisms are not easily testable, since disturbing the architecture of the tissue affects the correlation that is being inferred. A measure of the degree of confusion over the definition of growth factors has been revealed recently when it was concluded that '. . . Growth factors make cells grow. Growing cells makes more of everything, especially of ribosomes and the machinery of protein synthesis . . .' [19]. This view of what growth factors became differs significantly from that which was initially claimed, that is, they were specific signals that made their target cells enter the cycle. The changing, ephemeral, definitions of what growth factors are suggest a basic misunderstanding of their physiological worth in the context of the control of cell proliferation.

An alternative strategy towards discovering the role EGF plays in animals was used under the assumption that investigating EGF receptors in target cells may provide useful hints. Interestingly, the cell line that has the highest concentration of EGF receptors appears to be insensitive to a proliferative effect by EGF [20]. By virtue of having a high concentration of EGF binding sites, the human epidermoid carcinoma A-431 cell line became a preferred experimental model; curiously, however, EGF administration results in a clear and reproducible lowering of cell yields.

The advent of targeted gene disruption allowed for the generation of homozygous mice carrying the null mutation of growth factors and/or their receptors. In 1995, three groups reported that mice in which the EGF receptor gene was 'knocked-out' (EGFRKO mice) showed abnormal phenotypes [21–23]. The genetic background of mice used in these experiments influenced the severity of the adverse effects ranging from lethality at implantation to survival for up to 3 weeks after birth. The common features reported in the three papers were, opened eyelids at birth, thinner skin and wavy hair. Despite a thinner epidermis, Threadgill *et al.* found that the DNA synthesis in epidermal cells at birth was comparable to that of intact control mice [21]. Hair follicles were less abundant in one study [23] and normal in the other two. Impaired development of the hair follicle with premature expression of keratin was reported in one of the studies. Lung maturation was delayed, probably due to general growth retardation. Obviously, epithelial cell proliferation was not significantly impaired since the EGFRKO animals were of normal size at birth. While one study found developmental anomalies in the brain due to impaired migration and survival of nerve cells, another report found no brain anomalies [22]. Remarkably, the skin of EGFRKO mice transplanted into susceptible hosts proliferated more actively than the control grafts [24]. In summary, the lack of expression of the EGF receptor seems to affect several functions, including cell migration, the control of expression of genes involved in the development of skin appendages, and the development of the spongiotrophoblast in the placenta. Contrary to the claim in its name, however, EGF does not seem to play an obvious role on the proliferation of keratinocytes in these knockout mice. This is consistent with the data collected in cell culture experiments mentioned above. Other 'growth factors' such as TGF-α and keratinocyte growth factor (KGF) were also postulated to regulate epidermal proliferation. However, TGF-α and KGF 'knockout' mice exhibited normal epidermal growth. Moreover, double knockout mice (null for both growth factors) showed no evidence of altered epidermal proliferation or wound healing [25].

The many other putative growth factors that have been described suffer from comparable conceptual pitfalls to those of EGF, TGF-α, and KGF. For example, the fibroblast growth factors (FGFs) have been claimed to stimulate the proliferation of cells of mesodermal origin. However, FGFs inhibit cell proliferation in the epiphysial growth plate [26, 27]. The proliferation of these cells is responsible for the increased length of long bones. Moreover, 'gain-of-function' mutations in the FGF receptor that would lead to its constitutive activation result in human dwarfism syndromes because of early closing of the epiphysial plates. For example, achondroplasia is linked to mutations of the FGF receptor-3 (FGFR-3) [26]. Other mutations of this receptor are associated with thanatophoric dysplasia, a lethal form of dwarfism [28]. Finally, mutations of the FGFR-1 are linked to Pfeiffer syndrome characterized by premature closure of cranial sutures and limb defects [29]. These findings further undermine the notion that growth factors have a direct, proliferative, physiological role on their target cells.

#### *Oncogenes*

Oncogenes will be discussed in Part II, Chapter 8 when dealing with carcinogenesis.

### *Inhibitory factors within the positive hypothesis of control of cell proliferation*

Supporters of the positive hypothesis introduce inhibitory factors on an *ad hoc* basis to reconcile the paradox posed by the following observations: (a) putative growth factors are present in circulation or in extracellular spaces at levels that are apparently constant (no reference has been found indicating any type of temporal variation of short- or long-term duration); (b) the presence of plasma membrane receptors for these growth factors is ubiquitous; and (c) despite (a) and (b), a significant number of cell types remain quiescent in the intact animal. Textbooks and reviews on the subject of the control of cell proliferation acknowledge that inhibitory factors ought to play an important role in the regulation of cell proliferation both in intact animals and *in culture* conditions. However, inhibitors of cell proliferation are represented as the counterweight of growth factors in a scheme conforming to the metaphysical notion of yin/yang as proposed by Pardee [30].

Johnson summarized the properties of a series of negative regulators of cell proliferation [31]. No inhibitor categorized as 'surface inhibitors of cell cycling' [like TGF-β, mammastatin, glial maturating factor-beta, mammary-derived growth inhibitor (MDGI), contactinhibin, mouse brain cell surface glycopeptide (BCSG), sialoglycopeptide inhibitor (CeReS-18) and interferons] has shown a narrow target cell-specificity compatible with a hint of physiological relevance in controlling cell proliferation. All of these inhibitors have been extracted from arbitrarily selected organs or tissues like bovine mammary gland, ascitic fluid of Ehrlich mammary carcinoma cells, the brain of mice and cows and in the case of TGF-β, it was first characterized from tumor cells. An ever-increasing family of TGF-β-like molecules was shown to be present 'in a wide variety of fetal and adult tissues'; this challenges the specificity of these factors and highlights the *ad hoc* quality of the arguments raised in this field. The 'growth arrest-specific gene products' are derived from using subtracted libraries of NIH 3T3 cells directed at identifying genes whose activity correlate with cell cycle stages [32]. For instance, genes called gas 1 and gas 2 are down-regulated when serum is added to serum-starved, quiescent NIH 3T3 cell cultures. No physiological state is comparable to this strictly *in-culture* subterfuge to make cells appear quiescent. Johnson correctly points out the difficulty in assigning a reliable role to these genes in bringing to a halt the cell cycle during any of its phases.

### *Transforming growth factor beta (TGF-β)*

It would be difficult to critically and briefly evaluate the merits of the enormous body of data generated in the last two decades on TGF-β. Therefore, we have opted to quote *verbatim* the opinions of those who introduced TGF-β into the field of cell biology for the readership to form their own opinion on the subject.

The original communication by DeLarco and Todaro describing this molecule concluded that a new class of polypeptide trophic factors conferred properties associated with the transformed phenotype to fibroblasts *in culture* [33]. These factors purified from supernatant fluids of Moloney MuSV-transformed mouse 3T3 cells increased thymidine incorporation seven-fold in these cells over that of the untreated control. In 1981, Roberts *et al.* purified, from normal tissues, TGFs comparable to those previously purified from neoplastic tissues [34, 35]. In their words, these TGFs were 'potent inducers of anchorage-independent growth of untransformed, non-neoplastic, indicator cells', a property lacking in 'other known mitogenic growth factors such as insulin-like growth factors, platelet-derived growth factor, nerve growth factor, fibroblast growth factor, and EGF.' Meanwhile, also in 1980, Robert Holley, a Nobel laureate for the discovery of tRNA, reported that he had identified an inhibitor of cell proliferation from the supernatant of BSC-1 (African green monkey epithelial) cells. The data showed that while 78% of the cells were in $G_1$ in the absence of the inhibitor, the percentage increased to 93% when the inhibitory fraction was added. By counting cells over time, however, a vanishingly small inhibition was recorded. Later on, in 1984, Holley and Moses showed that TGF-β purified from outdated human platelets 'stimulated . . . growth of AKR-2B cells in soft-agar . . . and . . . inhibited DNA synthesis in AKR-2B, BSC-1 and CCL-64 (mink lung) cells.' In 1985, Roberts *et al.* concluded that TGF-β was the first peptide chalone purified to homogeneity and speculated that it was an autocrine inhibitor with implications regarding the genesis of cancer [36]. A year later, Like and Massague confirmed that Mv1Lu mink lung cells were inhibited by 0.3 nM TGF-β. These authors also remarked that Mv1Lu cells represented an ideal system to study the molecular basis of growth inhibition by TGF-β and that inhibition of the proliferative response to TGF-β occurs 'at a level distal from the receptors for growth-activating factors' [37].

In the last decade, work on TGF-β has been increasing at an exponential rate. There are several isoforms of this molecule; however, based on historical concerns, we will analyze the role(s) of TGF-β1 because this 25 kDa molecular mass homodimer was the one on which an inhibitory effect was originally reported. According to Sporn and Roberts, ' . . . a rigorous structural definition of What is TGF-β? is complicated by its potential for heterodimer formation . . . ' [38]. Sporn and Roberts addressed the functional role of TGF-β by stating:

> "The problem of a functional definition for TGF-β is much more complicated, perhaps almost impossible . . . As we noted before (Sporn and Roberts, 1988, 1990), in reality there is no such thing as a peptide 'growth factor'; all such molecules are multifunctional agents. This is particularly true for TGF-β, which may be considered the prototypical multifunctional signaling molecule (Roberts *et al.*, 1985). As is the case for all the other peptide growth factors, it is an element of a complex biological signaling language, providing the basis for intercellular (and perhaps even intracellular) communication in higher organisms. Thus, like a symbol or a letter of the alphabet in a language or code, the meaning of the action of TGF-β can only be considered in a cellular context. TGF-β always acts as a member of a set of other signals, and to understand its action one must always consider the biological context in which it acts . . ."

Finally, to have a better idea about what those who have championed TGF-β as a 'potent antiproliferative signal for many cell types' propose as the role of this molecule, we quote another short segment of their views: 'The function of TGF-β then is not to have an intrinsic action but to serve as a mechanism for coupling a cell to its environment, so that the cell has the plasticity to respond appropriately to changes in its environment or to changes in its own state.'

We have verified that TGF-β, used at the recommended concentrations, indeed inhibited the proliferation rate of those mink lung cells (unpublished data). However, when the susceptibility of other cell lines was tested, there was no inhibitory effect. It is puzzling to account for the inhibitory effect of TGF-β on mink lung cells *in culture*. Discounting a toxic effect (cells do not appear as intoxicated), it still remains unclear why TGF-β would be a specific inhibitor among, of all possible cells, these mink lung cells.

The fact that TGF-β does not affect the proliferative rate of other cells does not invalidate the possibility that TGF-β may play a role other than that of affecting cell proliferation. For instance, TGF-β is a homolog of the product of *decapentaplegic*, a *Drosophila* gene involved in the formation of dorsal–ventral polarity; this function does not involve an obvious inhibitory effect on cell proliferation [39]. By now a family of TGF-β molecules has been described which affects a variety of developmental functions; whether or not some of them play a role in regulating cell proliferation directly has not yet been made clear. However, TGF-β1-deficient animals (knockouts) are indistinguishable from their normal littermates at birth, and they appear normal for the first 2 weeks of life; this suggests that cell proliferation control is not affected. After this period, a number of disturbances of the immune system appeared [40, 41].

Because this book is aimed at readers that include non-specialists, we will not further belabor the same points on the many putative growth inhibitory factors described. They all share comparable pitfalls to those of TGF-β.

### *How are growth factors viewed in the context of the negative control hypothesis?*

The negative control hypothesis acknowledges that growth factors play a role as metabolic regulators (insulin, insulin-like growth factors), spreading factors for cells in culture (EGF), survival factors (colony-stimulating factors, EPO), and morphogens (fibroblast-growth factors, TGF-β). Contrary to inferences stemming from the adoption of *quiescence* as the default state, the available data argue against the notion that growth factors are direct stimulatory signals for the control of cell proliferation.

## Negative hypotheses

The negative hypotheses for the control of cell proliferation is based on the following premises: (a) *proliferation* is the default state of all living cells; (b) cells will exercise their constitutive, built-in capacity to proliferate when adequate nutritional requirements are met; and (c) cells will proliferate when extracellular or intracellular inhibitors are absent, or their down-stream effectors are inactive.

### *Inhibitory factors*

Chapter 5 will be dedicated to our experience with the inhibition of the proliferation of sex steroid target cells. Here, we will deal with *chalones*, putative inhibitors of cell proliferation whose existence was promoted four decades ago, and with inhibitory controls mediated by cell-to-cell communication.

#### *Chalones*

The concept of chalones was introduced by Bullough in 1962 [42]. This notion was based on the ideas developed by Weiss and Kavanau [3], and further explored by Iversen [43]. The concept was conceived in order to address the question: How is tissue renewal regulated in adult organisms? In particular, how is it that the epidermis responds to local events by altering its rate of renewal? These questions address events occurring at the tissue/organ hierarchical level, rather than at the cellular level. Therefore, the central issue here is, for example, how the skin in a certain region maintains its thickness. This issue is conceptually different from that of how cell proliferation is controlled (i.e. which is the default state of cells?, etc.). Iversen defined chalones as naturally occurring, physiological, cell-specific inhibitors of proliferation that are selectively produced by the tissue on which they act; their action is reversible [44]. Chalones would operate when there is no true quiescent state; therefore, the regulation of proliferative activity would take place within the cell cycle. Hence, chalones would act by prolonging the traverse of the $G_1$ or $G_2$ phases ($G_1$ and $G_2$ chalones). While accepting that the default state is *proliferation*, the chalone hypothesis accommodates the presence of both positive and negative signals; however, negative control becomes more important than positive control in a system whereby the default state is *proliferation*.

Bullough also subscribed to the existence of hormones that directly stimulate 'mitosis' [45]. This admission practically eliminated the chalone option as a serious challenge to the prevailing paradigm of positive controls. In the late 1980s, Iversen and colleagues described the isolation of several putative chalones [46]. These chalones are oligopeptides that have been thoroughly characterized biochemically; however, the small magnitude of the inhibitory effects obtained in animal experiments did not favor the acceptance of these molecules as mediators of the control of cell proliferation. On the one hand, it is likely that the effects produced by these chalones in tissues of fast turnover, such as the epidermis, should be of small magnitude, rather than the massive effects seen in the regenerating liver after partial hepatectomy, or in the prostate of castrated rats after prolonged (72 hours and longer) androgen administration. On the other hand, the claims by those who adopted the positive control hypothesis are not much more impressive; hence, it is clear that the acceptance of the latter over the former is based on *a priori* paradigm preferences, rather than on the strength of the supporting evidence.

### *Inhibitory controls mediated by cell-to-cell communication*

In a multicellular organism, the constitutive ability of cells to undergo proliferation may be curtailed by multiple means. That cell-to-cell interactions mediate the control of cell proliferation is illustrated by the regulation of the cell number in imaginal discs of *Drosophila*. The ablation of a portion of one imaginal disc is followed by localized cell proliferation at the site of the wound. Proliferation ceases when the cell number in the imaginal disc is regenerated to values before the substraction took place. Thus, regulation is intrinsic to the disc, and is controlled by positional information involving interactions between neighboring cells within the epithelium. The signal to stop proliferation when the imaginal disc achieves its full size is the establishment of coupling via gap junctions [47]. Developmental mutants that result in overgrowth and neoplasia of imaginal discs have been used to further understand intrinsic controls. In the *dco* mutant, the abnormal proliferative pattern seems to be caused by defective gap–junctional complexes [48], while the *fat* mutant disc cells lack zonulae adherents. The *fat* gene codes for a large relative of cadherins, plasma membrane-associated proteins that are in contact with each other in a tissue setting [49].

A role of cell adhesion molecules such as P-cadherin on the control of cell proliferation is suggested by lesions found in mice that are homozygous for a null mutation of the P-cadherin gene. This cadherin is expressed in the myoepithelial cells that are placed between the basement membrane and the lining epithelium of the mammary gland ducts. In females where the P-cadherin gene was not expressed, mammary development occurred earlier than in their wild-type counterparts. Also, the mammary glands of adult animals showed hyperplasia and dysplasia of the luminal epithelium. This suggests that the myoepithelial cells restrict the proliferation of adjacent epithelial cells by cell-to-cell contacts mediated by P-cadherin [50]. The study of cell-to-cell communication complements the efforts of those who study the control of cell proliferation through endocrine and paracrine mechanisms.

### *Anti-oncogenes or suppressor genes*

These concepts were developed within the context of carcinogenesis. Therefore, they will be dealt with in Part II, Chapter 9.

## Conclusions

The positive and negative hypotheses on the control of cell proliferation are based on mutually exclusive premises about the default state of cells in metazoa. When adopting *quiescence* as the default state, the control of cell proliferation requires positive signals or growth factors. When adopting *proliferation* as the default state, there is no need to postulate the existence of stimulatory factors that would impel cells to enter the cycle. Once it is acknowledged that the data gathered from the positive control hypothesis perspective do not support the notion that the growth factors characterized so far are the ultimate signals that move quiescent cells into proliferation, there are two avenues to pursue, namely, to continue looking for such evidence, or to switch paradigms.

We opted for switching paradigms based on our analysis of the conservation of the default state throughout evolution. From this perspective, the regulation of cell proliferation occurs through the exclusive effect of inhibitory factors. These factors are likely to act through endocrine, paracrine, and cell-to-cell mechanisms. Because the notion that *proliferation* is the default state for all cells has been mostly overlooked, most of these inhibitors have yet to be characterized physiologically. In the next chapter we will describe our experience with one such inhibitor.

By relieving growth factors from the role of positive levers for cell proliferation, attention should turn to alternative roles that these macromolecules may play. The data indicate that growth factors play an important regulatory role in the control of gene expression during development in metazoa (as morphogens that affect tissue inductions, axial plans, etc.).

As discussed in Chapter 2, at this time it is impossible to verify experimentally which is in fact the default state of cells in metazoa because of our lack of knowledge over the nutritional needs of cells. Hence, until this important issue is resolved the choice of the premises under which experiments are conducted in this area of research remains arbitrary. In previous chapters and in the current one we advanced arguments favoring the choice of *proliferation* as the most likely default state for all cells.

## References


1. **Michalopoulos, G.K. and DeFrances, M.C.** (1997) Liver regeneration. *Science* **276**: 60–66.

2. **Bard, J.B.L.** (1978) A quantitative model of liver regeneration in the rat. *J. Theor. Biol.* **73**: 509–530.

3. **Weiss, P. and Kavanau, J.L.** (1957) A model of growth control in mathematical terms. *J. Gen. Physiol.* **41**: 1–47.

4. **Uehara, Y., Minowa, O., Mori, C., Shiota, K., Kuno, J., Noda, T. and Kitamura, N.** (1995) Placental defect and embryonic lethality in mice lacking hepatocyte growth factor/scatter factor. *Nature* **373**: 702–705.

5. **Schmidt, C., Bladt, F., Goedecke, S., Brinkmann, V., Zschiesche, W., Sharpe, M. and Birchmeier, C.** (1995) Scatter factor/hepatocyte growth factor is essential for liver development. *Nature* **373**: 699–702.

6. **Columbano, A. and Shinozuka, H.** (1996) Liver regeneration versus direct hyperplasia. *FASEB* **10**: 1118–1128.

7. **LaBrecque, D.** (1994) Liver regeneration: a picture emerges for the puzzle. *Am. J. Gastroenterol.* **89**: S86–S96.

8. **Cherington, P.V.** (1984) *Mammalian Cell Culture: The Use of Serum-free Hormone-supplemented Media.* Plenum Press, New York, pp. 17–52.

9. **Bissell, M.J.** (1981) The differentiated state of normal and malignant cells or how to define a normal cell in culture. *Int. Rev. Cytol.* **70**: 27–100.

10. **Alberts, B., Bray, D., Lewis, J.G., Raff, M., Roberts, K. and Watson, J.D.** (1994) *Molecular Biology of the Cell*, 3rd Edn. Garland Publishing, New York, p. 159.

11. **Gospodarowicz, D. and Moran, J.S.** (1976) Growth factors in mammalian cell culture. *Annu. Rev. Biochem.* **45**: 531–558.

12. **Baserga, R.** (1985) *The Biology of Cell Reproduction.* Harvard University Press, Cambridge, MA, p. 118.

13. **Durum, S.K. and Muegge, K.** (1998) *Cytokine Knockouts.* Humana Press, Totowa, NJ, pp. VII–XVI.

14. **Gilbert, S.** (1997) *Developmental Biology*, 5th Edn. Sinauer Associates, Sunderland, MA, pp. 657–663.

15. **Deshmukh, M. and Johnson, E.M.** (1997) Programmed cell death in neurons: focus on the pathway of nerve growth factor deprivation-induced death of sympathic neurons. *Mol. Pharmacol.* **51**: 897–906.

16. **Torcia, M., Bracci-Laudiero, L., Lucibello, M., Nencioni, L., Labardi, D., Rubartelli, A., Cozzolino, F., Aloe, L. and Garaci, E.** (1996) Nerve growth factor is an autocrine survival factor for memory B lymphocytes. *Cell* **85**: 345–356.

17. **Cohen, S.** (1965) Growth factors and morphogenic induction. In: *Developmental and Metabolic Control Mechanisms and Neoplasia*. Williams and Wilkins, Baltimore, MD, pp. 251–272.

18. **Barrandon, Y. and Green, H.** (1987) Cell migration is essential for sustained growth of keratinocyte colonies: the roles of transforming growth factor alpha and epidermal growth factor. *Cell* **50**: 1131–1137.

19. **Hunt, T. and Nasmyth, K.** (1997) Cell multiplication. *Curr. Opinion Cell Biol.* **9**: 765–767.

20. **Kawamoto, T., Sato, J.D., Le, A., Polikoff, J., Sato, G.H. and Mendelsohn, J.** (1983) Growth stimulation of A431 cells by epidermal growth factor: identification of high-affinity receptor monoclonal antibody. *Proc. Natl Acad. Sci. USA* **80**: 1337–1341.

21. **Threadgill, D.W., Dlugosz, A.A., Hansen, L.A., Tennenbaum, T., Lichti, U., Yee, D., LaMantia, C., Mourton, T., Herrup, K., Harris, R.C., Barnard, J.A., Yuspa, S.H., Coffey, R.J. and Magnuson, T.** (1995) Targeted disruption of mouse EGF receptor: effect of genetic background on mutant phenotype. *Science* **269**: 230–234.

22. **Sibilia, M. and Wagner, E.F.** (1995) Strain-dependent epithelial defects in mice lacking the EGF receptor. *Science* **269**: 234–238.

23. **Miettinen, P.J., Berger, J.E., Meneses, J., Phung, Y., Pedersen, R.A., Werb, Z. and Derynck, R.** (1995) Epithelial immaturity and multiorgan failure in mice lacking epidermal growth factor receptor. *Nature* **376**: 337–341.

24. **Hansen, L.A., Alexander, N., Hogan, M.E., Sundberg, J.P., Dlugosz, A., Threadgill, D.W., Magnuson, T. and Yuspa, S.H.** (1997) Genetically null mice reveal a central role for epidermal growth factor receptor in the differentiation of the hair follicle and normal hair development. *Am. J. Pathol.* **150**: 1959–1975.

25. **Guo, L., Degenstein, L. and Fuchs, E.** (1996) Keratinocyte growth factor is required for hair development but not for wound healing. *Genes & Develop.* **10**: 165–175.

26. **Webster, M.K. and Donoughue, D.J.** (1996) Constitutive activation of fibroblast growth factor receptor-3 by the transmembrane domain point mutation found in achondroplasia. *EMBO J.* **15**: 520–527.

27. **Deng, C., Wynshaw-Boris, A., Zhou, F., Kuo, A. and Leder, P.** (1996) Fibroblast growth factor receptor-3 is a negative regulator of bone growth. *Cell* **84**: 911–921.

28. **Bellus, G.A., McIntosh, I., Smith, E.A., Aylsworth, A.S., Kaitila, I., Horton, W.A., Greenhaw, G.A., Hecht, J.T. and Francomano, C.A.** (1995) A recurrent mutation in the tyrosine kinase domain of fibroblast growth factor receptor-3 causes hypochondroplasia. *Nature Genet.* **10**: 357–359.

29. **Goldfarb, M.P.** (1996) Functions of fibroblast growth factors in vertebrate development. *Cytokine Growth Factor Rev.* **7**: 311–325.

30. **Pardee, A.B.** (1987) The Yang and Yin of cell proliferation: an overview. *J. Cell Physiol.* **5**(Suppl): 107–110.

31. **Johnson, T.** (1994) Negative regulators of cell proliferation. *Pharmac. Ther.* **62**: 247–265.

32. **Schneider, C., King, R.M. and Philipson, L.** (1988) Genes specifically expressed at growth arrest of mammalian cells. *Cell* **54**: 787–793.

33. **deLarco, J.E. and Todaro, G.J.** (1978) Growth factors from murine sarcoma virus-transformed cells. *Proc. Natl Acad. Sci. USA* **75**: 4001–4005.

34. **Roberts, A.B., Anzano, M.A., Lamb, L.C., Smith, J.M. and Sporn, M.B.** (1981) New class of transforming growth factors potentiated by epidermal growth factor: isolation from non-neoplastic tissues. *Proc. Natl Acad. Sci. USA* **78**: 5339–5343.

35. **Roberts, A.B., Lamb, L.C., Newton, D.L., Sporn, M.B., deLarco, J.E. and Todaro, G.J.** (1980) Transforming growth factors: isolation of polypeptides from virally and chemically transformed cells by acid/ethanol extraction. *Proc. Natl Acad. Sci. USA* **77**: 3494–3498.

36. **Roberts, A.B., Anzano, M.A., Wakefield, L.M., Roche, N.S., Stern, D.F. and Sporn, M.B.** (1985) Type beta transforming growth factor: a bifunctional regulator of cellular growth. *Proc. Natl Acad. Sci. USA* **82**: 119–123.

37. **Like, B. and Massague, J.** (1986) The antiproliferative effect of type beta transforming growth factor occurs at a level distal from receptors for growth-activating factors. *J. Biol. Chem.* **261**: 13426–13429.

38. **Sporn, M.B. and Roberts, A.B.** (1990) TGF-beta: problems and prospects. *Cell Regulation* **1**: 875–882.

39. **Gelbart, W.M.** (1989) The decapentaplegic gene: a TGF-beta homologue controlling pattern formation in Drosophila. *Development* **107**(Suppl): 65–74.

40. **Kallapur, S., Shull, M. and Doetschman, T.** (1998) Phenotypes of TGF-beta knockout mice. In: *Cytokine Knockouts* (eds S.K. Durum and K. Muegge). Humana Press, Totowa, NJ, pp. 335–368.

41. **Kulkarni, A.B. and Letterio, J.J.** (1998) The transforming growth factor-beta1 knockout mouse. In: *Cytokine Knockouts* (eds S.K. Durum and K. Muegge). Humana Press, Totowa, NJ, pp. 369–400.

42. **Bullough, W.S.** (1962) The control of mitotic activity in adult mammalian tissues. *Biol. Rev.* **37**: 307–342.

43. **Iversen, O.H.** (1961) The regulation of cell numbers in epidermis: a cybernetic point of view. *Acta Pathol.* **148**: 91–96.

44. **Iversen, O.H.** (1981) The chalones. In: *Handbook of Experimental Pharmacology* (ed. R. Baserga). Springer-Verlag, Berlin, pp. 491–550.

45. **Bullough, W.S.** (1977) Chalones and cancer. In: *Growth Kinetics and Biochemical Regulation of Normal and Malignant Cells* (eds B. Drewinko and R.M. Humphries). Williams & Wilkins, New York, pp. 77–89.

46. **Elgjo, K.** (1993) Pentapeptide growth inhibitors, carcinogenesis, and cancer. In: *New Frontiers in Cancer Causation* (ed. O.H. Iversen). Taylor & Francis, Washington, DC, pp. 125–137.

47. **Bryant, P.J. and Schmidt, O.** (1990) The genetic control of cell proliferation in Drosophila imaginal discs. *J. Cell Sci.* **13**(Suppl): 169–189.

48. **Jursnich, V.A., Fraser, S.E., Held, L.I., Ryerse, J. and Bryant, P.J.** (1990) Defective gap-junctional communication associated with imaginal disc overgrowth and degneration caused by mutations of the dco gene in Drosophila. *Dev. Biol.* **140**: 413–429.

49. **Watson, K.L., Justice, R.W. and Bryant, P.J.** (1994) Drosophila in cancer research: the first fifty tumor suppressor genes. *J. Cell Sci.* **18**(Suppl): 19–33.

50. **Radice, G.L., Ferreira-Cornwell, M.C., Robinson, S.D., Rayburn, H., Chodosh, L.A., Takeichi, M. and Hynes, R.O.** (1997) Precocious mammary gland development in P-cadherin-deficient mice. *J. Cell Biol.* **139**: 1025–1032.

## Chapter 5

# Sex Hormone-mediated Control of Cell Proliferation

"It is also a good rule not to put too much confidence in the observational results that are put forward *until they are confirmed by theory.*"

*Sir Arthur Eddington (1934), quoted in: Horace F. Judson (1979) The Eighth Day of Creation. Penguin Books, London, p. 93*

". . . both the ideas that science generates and the way in which science is carried out are entirely counter-intuitive and against common sense. . . . Science does not fit with our natural expectations."

*Lewis Wolpert (1992) The Unnatural Nature of Science. Faber & Faber, London, p. 1*

## Introduction

The study of the control of cell proliferation in metazoa requires the selection of a suitable experimental model. In mammalian species, the cells of the lining of the accessory sex organs appeared to be the most promising tool. Several compelling reasons sustained this selection. First, sex hormones are acknowledged to be physiologically relevant, highly specific signals that affect the proliferation and trophism of their target organs and cells. The surgical extirpation of the ovaries or the testes (gonadectomy) in adult mammals is followed by atrophy of the accessory sex organs. Administration of estrogens in females and androgens in males is followed by a massive proliferative response in the epithelial lining of their respective target organs. Second, sex steroids reach all tissues of the body, but only the cells that have specific receptors are able to recognize them as *signals*. Finally, always following the inevitable cartesian method, 'established' target cell lines mimic many of the properties of their 'normal' counterparts, and they can be studied *in culture*, circumventing the many regulatory controls operating in the whole organism. Results from these experiments helped us in formulating hypotheses on the control of cell proliferation in the animal.

## Control of cell proliferation by estrogens

In rodents, the lining of the uterus follows cyclical changes characterized by an increase in the number (hyperplasia) and the size (hypertrophy) of epithelial cells, followed by cell death (apoptosis). The **proximate** cause of these changes is the variation of blood estrogen levels during the 4–5-day-long ovarian cycle. Histological and physiological changes occur in several estrogen target cells (pituitary, mammary glands, etc.); however, we will focus our analysis on changes witnessed in the epithelial lining of the uterus (endometrium) and vagina. During proestrus, shortly after estrogen plasma levels increase, there is an increased proliferative activity of these epithelia. After ovulation, the proliferative activity ceases, due to the action of another ovarian hormone, progesterone. Ovariectomy is followed by cell death in the endometrium; the remaining cells in the lining epithelium now appear small and rarely proliferate. Administration of a single dose of estrogen to ovariectomized mice results in a cessation of cell death, followed by two cycles of cell proliferation in the uterine epithelium lining [1]. The first wave results in a doubling of the cell number, which is followed by a second wave that affects a smaller number of cells. Finally, this is followed by the re-establishment of the characteristics prevalent after ovariectomy.

However, if instead of a single estrogen dose, animals are given estrogens over a prolonged period of time (let us say, 1 week), this steady level of estrogens induces a short-lived proliferative response, which is followed by a prolonged proliferative quiescence (proliferative shutoff) (*Figure 5.1*). That is, steady estrogen levels induce a biphasic response: first, proliferation, and next, inhibition of cell proliferation [2].

### *Estrogen receptors*

Until momentous experiments by Jensen and Jacobson in the 1950s, little was known about the fate of estrogens in the organism [3]. Their experiments revealed that the administration of radiolabeled estradiol to rodents resulted in its accumulation for a few hours in the uterus and other target organs. Remarkably, the hormone disappeared from these organs long before the signs of proliferative activity were manifested, that is, before cells entered the S phase of the cell cycle. This experiment led to the discovery of estrogen receptors, the proteins responsible for the retention of estradiol inside target cells, and resulted in a large-scale

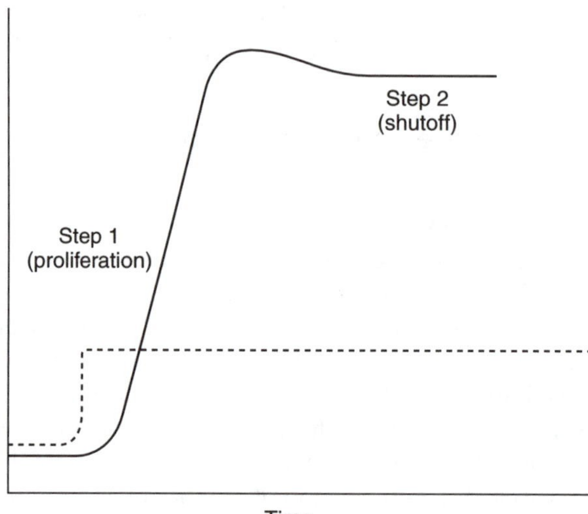


**Figure 5.1.** Schematic representation of the control of cell proliferation by estrogens and androgens. In the absence of sex hormones (by gonadectomy), these target cells are prevented from proliferating. Step 1: a single injection of the relevant sex hormone is followed by the proliferation of their specific target cells (e.g. uterine cells for estradiol, or prostate cells for androgens; ———, cell number). Step 2: if the hormone concentration (-----) is kept high for several days by implanting a pellet with the pertinent hormone, the proliferative response is followed by a period of proliferative arrest.

research program aimed at purifying the receptor and understanding its role as a mediator of estrogen action. It is now understood that estrogen receptors operate as transcription factors, which are necessary in the process of estrogen-induced gene expression. In the absence of estrogens, the receptors are inactive. Once the receptors bind to estrogen, they become 'activated'; that is, they interact with specific DNA sequences, and with other transcription factors. This, in turn, allows transcription of the specific genes into mRNAs which are next translated into specific proteins [4].

## Hypotheses

### *Direct-positive hypothesis. The advent of estrogen-target established cell lines*

In the 1950s, Furth developed a protocol whereby chronic administration of supraphysiological doses of estrogens to rats resulted in the development of pituitary tumors [5]. These tumors were transplantable and grew following estrogen administration, while they failed to grow or grew slowly in ovariectomized females and in males. Cell lines were established from these tumors; these cells retained their estrogen receptors and grew in a hormone-dependent fashion when inoculated into rats of the same inbred strain from which these cells originated (syngeneic rats). Our own involvement in this subject started with the establishment of estrogen-target pituitary cell lines. In the 1970s, all research programs on estrogen action on cell proliferation, including ours, were conducted under the implicit premise that the default state of metazoan cells was *quiescence*. Estrogens were supposed to directly stimulate the proliferation of their target cells, via the estrogen receptor, by inducing the entry of these cells into the cell cycle (direct-positive hypothesis, *Figure 5.2*). In other words, estrogens were postulated to be the *ultimate* positive signal that induced cell proliferation. Soon thereafter, we faced a paradox: namely, while these cells developed into estrogen-dependent tumors when inoculated into animals, *in culture* they proliferated at the same rate regardless of the addition of estrogens [6].

### *Indirect-positive hypothesis. Evaluation of cell proliferation as a discrete marker*

In the mid-1970s, we became convinced that estradiol was the *proximate* and not the *ultimate* cause of the proliferative event described above. This conclusion was based on the following facts: first, estradiol administration to animals did not increase cell proliferation in some of the normal tissues and neoplastic cells that carried estrogen receptors. This implied that the presence of estrogen receptors might have been necessary but not sufficient for the proliferative response to estradiol to occur. Second, neither our data nor those of other scientists showed a direct proliferative effect of estradiol on target cells *in culture*. Hence, we concluded that estradiol affected the proliferation of its target cells indirectly [7]. Hypothetically, there was an intermediary step requiring the intervention of growth factors secreted by estrogen target organs under estradiol stimulation [8, 9]. By following this rationale, researchers were adopting the *indirect-positive hypothesis* (*Figure 5.2*).

Three variations of this hypothesis were proposed: (a) the endocrine hypothesis, whereby the growth factors were produced in an intermediary organ and secreted into the bloodstream, (b) the paracrine hypothesis, claiming that the growth factors were secreted by cells in an adjacent tissue or cell type, and (c) the autocrine hypothesis, which postulated that the growth factors were secreted by the same cells that proliferated upon estrogen administration [10]. In 1980, we analyzed the data produced in support of these hypotheses in a guest editorial of

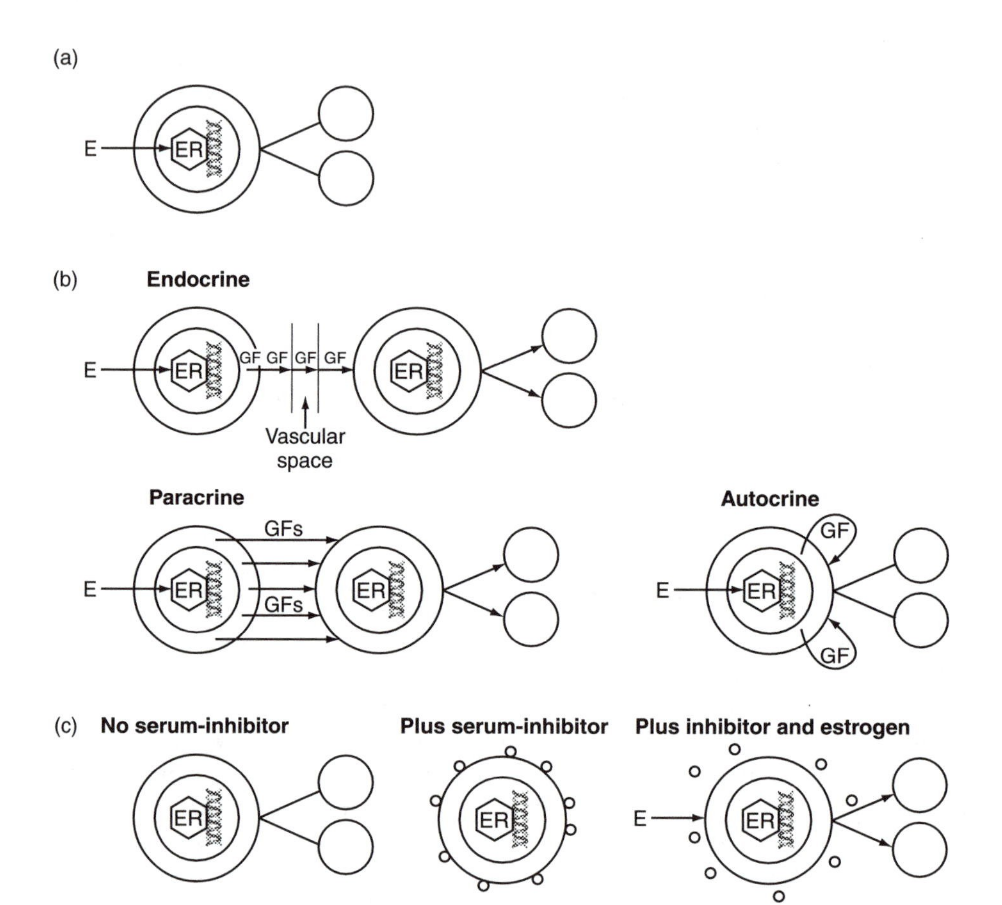


**Figure 5.2.** Schematic representation of the positive and negative hypotheses for the control of cell proliferation by estrogens. (a) The direct positive hypothesis assumes that the default state of cells in metazoa is *quiescence*. Under this hypothesis, estrogens would themselves stimulate the proliferation of their target cells (direct-positive hypothesis). (b) Alternatively, estrogens would stimulate the synthesis and/or secretion of growth factors which in turn would stimulate the proliferation of their target cells (indirect-positive hypothesis). (c) The indirect-negative hypothesis assumes that the default state of all cells is *proliferation*. In the absence of serum/plasma (*in culture* conditions only), estrogen-target cells proliferate. In the presence of serum/plasma (*in culture* or in animals) and in the absence of estrogens, these same cells are prevented from proliferating. These cells will proliferate when estrogens are administered; proliferation will ensue because of the neutralization of the serum/plasma-borne inhibitor (estrocolyone-I). This would represent an effective estrogen-mediated disinhibition of these quiescent estrogen-target cells.

the *Journal of the National Cancer Institute* [6]. There, we called attention to the fact that data claimed to demonstrate effects of growth factors and estrogens were not gathered by counting cells during the exponential phase of proliferation, a *sine qua non* for validity (see Chapter 3). Further, we suggested the rigorous application of the rule that cell proliferation should be estimated by measuring the doubling time of cell populations during the exponential phase of proliferation. In addition, the data we gathered on the role of the pituitary gland as a source of estrogen-induced growth factors, presumably acting on the uterus, were inconsistent with the indirect positive endocrine hypothesis. Estrogen treatment resulted in the increase of the uterine weight both in ovariectomized only, and in hypophysectomized and ovariectomized rats, however, the magnitude of the response was larger in the former than in the latter groups [9]. Estrogen-induced proliferative activity was observed in the uterine lining in both experimental conditions. This suggested that the pituitary gland did not have a crucial role in estrogen-mediated induction of cell proliferation. In sum, the overall data were inconclusive in regard to the positive hypothesis.

### *Indirect-negative hypothesis. Choosing the least traveled road*

It became plain to us that the above-mentioned paradox *in animal/in culture* could be resolved by just switching the premises. Namely, instead of adopting the premise that cells were quiescent waiting for the positive stimulus(i) to signal that they should enter the cycle, now one could postulate that cells were always ready to proliferate, and that they would do so unless a specific inhibitor(s) prevented them from expressing this constitutive, built-in ability. This was the origin of the *indirect-negative hypothesis* (*Figure 5.2*). Estradiol would increase the proliferation of its target cells by blocking either of the two following mechanisms. In the first, estradiol would neutralize the effect of a blood-borne inhibitor by binding to it or by blocking its action at the target cell level. In the second option, estradiol would inhibit the synthesis or release of the inhibitor at its source, and thus decrease its serum concentration. In either case, the outcome would be the proliferation of the estrogen-target cells.

Evidence supporting our proposed interpretation of the data also came from experiments showing that mouse vaginal epithelium explants proliferated regardless of the presence of estradiol in the serumless medium [11, 12]. These data, and the data from our own experiments, reinforced the notion that our postulated inhibitor was blood-borne. By switching to this new paradigm, we altogether abandoned the search for growth factors. It was at this point in our research program that we began to question the rationale underlying the premise that *quiescence* was the default state of cells in metazoa.

### *MCF7 cells: the foremost human breast estrogen-target established cell line*

We moved on to explore the role of serum-borne inhibitors on the proliferation of estrogen-target cells. The first estrogen-target cell line established from a breast cancer was called MCF7. This cell line was established at the Michigan Cancer Foundation in Detroit in the early 1970s. Cells from this line grew as tumors when inoculated into athymic nude mice (these mice tolerate grafts from species other than their own). Remarkably, tumors only developed at the site of inoculation when these mice were treated with estrogens. All these features convinced researchers in this field to adopt MCF7 cells as the most reliable *in-culture* model in which to study the mechanism of estrogen action, to the point that the estrogen receptor gene was isolated from these cells; so was the receptor protein. Still, a major obstacle to testing the effect of estrogens on the proliferation of MCF7 cells *in culture* was the presence of estrogens in serum. The traditional method of estrogen removal, charcoal-dextran stripping, was used by several laboratories. However, in spite of claims to the contrary, there was no significant difference on the proliferation rates of these cells either in the presence or absence of estradiol in medium supplemented with charcoal-dextran stripped serum. A reason for an apparent lack of a significant proliferative effect by estradiol was the presence of an estrogenic contaminant in some preparations of phenol-red, a pH indicator used in most culture medium formulations [13]. It was also subsequently shown that there is considerable variation in the ability of different sublines of MCF7 cells to become inhibited by estrogenless serum [14]. Once again, the devil is in the details.

### *There was indeed a plasma-borne inhibitor*

In 1983, for purification purposes, we adopted an experimental human model because human serum is a more abundant source of material than rat serum. We reasoned that if serum contained an inhibitor of the proliferation of estrogen-target cells, it will be more abundant in the serum of women at the beginning of the menstrual cycle (day 1), and in male donors. This line of thought followed the correlation between low mitotic index in target cells (uterus, vagina) and low circulating levels of estrogens. The putative inhibitor would be less effective or abundant in the serum of women at midcycle, when the levels of estradiol are high (days 14–15). Also, and most important, the degree of inhibition of MCF7 cell proliferation would be proportional to the amount of serum in the culture medium. We found that both predictions were met: the higher the level of serum from men, or from women on the first day of their cycle, the lower the number of cells in the dish at the end of the experiment (7–10 days). Serum from women at midcycle lacked inhibitory activity at all concentrations tested. Addition of estradiol reversed the inhibitory activity of serum from men and from women on the first day of their cycle. However, these experiments could not discern whether the lack of inhibition by serum from women at midcycle was due to low inhibitor levels or high estrogen levels. To resolve this point, we then removed estrogens from serum to discriminate between these options. Once made estrogenless, all serum samples were of comparable inhibitory potency, regardless of whether they were drawn from women at any stage of the menstrual cycle, or from men. The degree of inhibition was, again, proportional to the amount of serum in the medium. Estradiol reversed this inhibitory effect. These data favored the notion that estradiol either neutralized the level of the serum inhibitor by binding to it or blocked its action at the level of the estrogen-target cells. Following a time-honored tradition, we named this serum-borne inhibitor of the proliferation of estrogen-target cells with a shortened name. This inhibitor became known as estrocolyone-I (from the greek κολεω: to stop).

In the meantime, Sato's group had developed a serumless medium that allowed MCF7 cells to proliferate for a sufficiently prolonged period to measure proliferation rates [15]. We tried this medium but we omitted some components including EGF, as growth factors did not play a role in our hypothesis; to our delight the cells proliferated maximally in these conditions, regardless of whether or not estrogens were added. These experiments also suggested that serum contained a specific inhibitor of the proliferation of estrogen-target cells [16]. Soon after we published our findings, Lykkesfeldt and Briand at the Fibiger Institute in Copenhagen, Denmark, confirmed our results [17]. In addition, they developed an MCF7 cell variant that could be propagated in growth factor-free, serumless medium and which for over 10 years has maintained its ability to be inhibited whenever exposed to estrogenless serum-supplemented medium [18]. While we were developing our cell culture model, Bern and his collaborators generated conclusive evidence favoring the notion that mouse vaginal and uterine primary and organ cultures carrying functional estrogen receptors proliferated in serumless medium regardless of the presence of estradiol [19, 20].

In summary, data stemming from different *in-culture* experimental models and based on different working hypotheses pointed to the lack of a direct effect of estradiol on cell proliferation despite the presence of functional estrogen receptors [19, 21]. Moreover, these results were also consistent with the notion that *proliferation* was the default state of these cells [7]. Finally, advances in recombinant DNA technology allowed researchers to explore the effects of the estrogen receptor in cells that did not express it spontaneously, by transfecting such cells with an expression vector containing the estrogen receptor cDNA. Invariably, these transfected cells did not show an estradiol-mediated increased cell proliferation rate [22]. These data were consistent with our hypothesis that the ultimate control was mediated by a serum-borne inhibitor, and that estrogens could only increase proliferation in cells that are responsive to such inhibitor.

*The final identification of the plasma-borne inhibitor (estrocolyone-I)*

In the process of purifying estrocolyone-I, we consistently observed that the inhibitory activity appeared in purified fractions that contained serum albumin. Meanwhile, the Danish group published a paper claiming that commercial preparations of bovine serum albumin inhibited the proliferation of MCF7 cells and that estradiol reversed this inhibition [23], which we confirmed. However, these commercially available albumin preparations were contaminated with other proteins. This left open the possibility that one of the contaminants, rather than albumin, could be the inhibitor. To rule out albumin being contaminated with a molecule tightly bound to it in all serum-derived commercial preparations, we decided to test recombinant human albumin, devoid of serum contaminants. Recombinant albumin is produced by yeast cells grown in a synthetic medium devoid of animal proteins. Delta Biotechnology (Nottingham, UK) generously supplied us with recombinant human serum albumin; it inhibited the proliferation of estrogen-target cells in a comparable manner to that of serum-derived human albumin. Finally, we concluded that albumin was the long-sought estrocolyone-I, the inhibitor of the proliferation of estrogen-target cells [24].

Serum albumin is a protein characterized by a three domain structure that must have evolved from the duplication of an ancestral gene; these domains share common structural features, but have also unique properties. We tested three recombinant albumin fragments: Domain I alone, another protein encompassing Domain I and Domain II, and a protein encompassing the entire Domain III. Both Domain I and the Domains I and II fragments had inhibitory activity; the latter was a more potent inhibitor than the former. Domain III was devoid of inhibitory activity. In addition, when albumin was removed from serum, the remaining serum proteins lacked estrocolyone-I activity; thus, the inhibitory activity of serum was totally due to its albumin content [24].

In summary, the human breast MCF7 cells should be considered as *estrogen-target* cells (by virtue of their carrying estrogen receptors), and *serum (estrocolyone-I)-sensitive* because they are inhibited from proliferating when exposed to estrogenless serum. We expect that our findings *in culture* could be successfully extrapolated to the control mechanisms on the proliferation of estrogen-target, serum-sensitive cells in animals, including humans. This is currently being investigated in our laboratories.

### *And now . . . back to the animal*

While we were dedicated to characterizing the effect of recombinant albumin on MCF7 cells, our colleagues were exploring the indirect-positive hypotheses by using animal models. They applied the novel technology of gene inactivation to produce mice lacking the expression of estrogen and growth factor receptors. Their objective was to understand the role of these receptors in the control of cell proliferation and in stroma–epithelium interactions.

*Does EGF play a role in the proliferation of endometrial cells?*

Let us, for the sake of argument, explore the research program based on the premise that *quiescence* is the default state of cells in metazoa. Our colleagues found that the expression of growth factors (EGF and insulin-like growth factor) [25] and their receptors [26] were increased in target organs after estrogen administration in a temporal pattern consistent with their involvement in the proliferative response. In addition, EGF implants induced both cell proliferation **and** estrogen-regulated genes in uterine and vaginal epithelia in mice [27]. However, the notion that EGF had a direct proliferative effect on estrogen-target cells was contradicted by the lack of such an effect in mice whose estradiol receptors have been inactivated (estrogen receptor knockout, ERKO mice) [28]. Alternatively, the stimulatory effect of EGF implants in normal mice was postulated to occur through the estrogen receptor ('cross-talk') [29]. This would involve EGF binding to its receptor which, in turn, would produce a signal that would have activated estrogen receptors in the absence of estrogens. This would explain the lack of effect of EGF in ERKO mice. However, uterine tissue from animals lacking the EGF receptor respond normally to estrogens [30]; this implies that EGF is not a necessary component of the pathway that mediates estrogen action. In conclusion, these are conflicting data within the positive control paradigm. They highlight the fact that estrogen receptors are necessary for the proliferative effect of estradiol, while the EGF receptors are not.

*Is endometrial cell proliferation mediated by stroma–epithelial interactions?*

An alternative hypothesis based on the interaction between the stroma and the epithelia of estrogen-target cells was formulated to explain the induction of cell proliferation by sex hormones [31, 32]. It proposed that sex steroids act upon the stroma-target cells by inducing them to secrete growth factors that would, in turn, stimulate the proliferation of the adjacent epithelial cells. This notion has been explored by recombining epithelium and stroma and placing these 'tissue recombinants' in nude mice who would not reject them. Tissue recombinants work best when 'immature' animals are used as the source of stroma; that is, before extensive gland formation occurs as a result of epithelial cell invasion of the stroma. Hence, these epithelium–stroma interactions represent developmental events not evaluated when dealing with adult tissues. In other words, tissue recombination experiments represent the summation of morphogenetic events in addition to informational events taking place during adulthood. More to the point, they may not be comparable to experiments whereby estrogens or growth factors are injected into adult ovariectomized animals to evaluate the role of those molecules on the proliferation of endometrial cells. Endometrial epithelial cells from the ERKO mouse proliferated upon estrogen treatment when recombined with endometrial stroma from 'wild'-type mice, which contain normal estrogen receptors. This very interesting phenomenon, however, is difficult to integrate with other data which show that isolated epithelial uterine and vaginal cells proliferate *in culture* regardless of the presence of estrogens. Also, when the architecture of the organ, and consequently, the contact between stroma and epithelium remains intact *in culture*, epithelial cells proliferate in the absence of estrogen and growth factors [11, 12, 20]. If anything, these latter experiments indicate that inhibitory influences that exist *in animals* have been removed.

One way to reconcile the stroma–epithelium interaction data with data gathered in cell and organ culture models is to propose that there are two inhibitory pathways controlling the proliferation of epithelial cells in estrogen-target organs. The first would act directly on the epithelial cells (albumin inhibition, and its neutralization by estrogen through the estrogen receptor), and the other one would be mediated by inhibitory factors produced in stromal cells. To explain why it is that the stroma has ceased to send inhibitory signals to the epithelium in organ culture experiments, one may suggest that the synthesis of inhibitors by the stromal cells is induced by albumin and repressed by estrogens. Hence, in organ culture, the stromal cells would cease to secrete this inhibitor in the absence of albumin. Regardless of all these theoretical possibilities, one cannot ignore the powerful impact of seeing that cells just removed from the organism proliferate as soon as they are placed in culture conditions.

### *Can data gathered through different paradigms be integrated?*

Data based on conflicting paradigms can seldom be reconciled. This is because when researchers, explicitly or implicitly, adopt opposite premises they design experiments addressing different questions. For instance, when adopting the premise that *proliferation* is the default state of all cells, the question to be addressed when dealing with estrogen-target cells is 'How is it that albumin mediates the inhibitory response?' and 'How does the removal of albumin (or estrogen addition) result in the re-entry of these cells into the cell cycle?' Answering these questions entails finding out how target cells decode the message provided by albumin in order to stop cell proliferation. This strategy represents a significant departure from the positive control program, a program which searches for estradiol-induced genes that directly or indirectly regulate cell cycle traverse (cyclins and proto-oncogenes such as *fos*, *jun*, *myc*, etc.), and also searches for explanations of how 'growth factors' fit into the estrogen-mediated pathway. In other words, the negative control hypothesis seeks clarification over how a signal (in this case, albumin) prevents the target cells from entering the cycle. The removal of the inhibitory signal would allow the default state to be expressed (cells will proliferate). Removal of the inhibitor (albumin) would be equivalent to the addition of estrogens as seen by the cell. Whatever happens within the cycle is considered irrelevant to the control of cell proliferation because all cells perform such a cycle using similar machinery. To the contrary, the positive hypotheses based on the premise that the default state is *quiescence* assume that either estrogens or estrogen-induced growth factors would affect the expression of genes that code for the cycle effectors (cyclins, proto-oncogenes) through their respective receptors. Hence, significant portions of the data gathered by either research program are irrelevant to the other competing program. Finally, the resolution of these important issues may have to wait until the *Zeitgeist* changes . . . and/or the 'old philosophers' die (see Chapter 1). Less often, conflicts are settled by new data.

### *Why is it that prolonged estrogen treatment is not followed by a steady proliferation of cells in the uterus?*

Even if estradiol cancels the inhibitory effect of albumin, it remains to be explained why a massive hyperplastic reaction in the uterine epithelium of rodents does not follow the continued administration of estradiol. A plausible answer was provided by experiments published over two decades ago by Gorski's group. They showed that uterine cells subjected to repeated estradiol injections would, in their words, enter a period of proliferative 'shutoff' [2]. In this 'shutoff' step, estradiol, having entered the cell and become bound to its receptor in the nucleus, would induce the transcription of a gene(s) whose effect(s) would be to inhibit their target cells from entering a new cell cycle. This fail-safe combination would prevent an iterative proliferation of these cells and, therefore, the risk of hyperplasia, dysplasia, and subsequent stages leading to neoplasia.

## Control of cell proliferation by androgens

The sustained administration of androgens to castrated male rats results in three distinct effects on the epithelial cells of the prostate gland: inhibition of cell death, increase of the proliferation rate, and once the gland reaches adult size, inhibition of cell proliferation [33]. These effects are comparable to those elicited by estrogens on the epithelial lining of the endometrium.

The evidence collected in cell culture when adopting the premise that *proliferation* is the default state of all cells is consistent with the notion that androgens affect the proliferation of their target cells in animals [33] and *in culture*, in a manner comparable to that regulated by estrogens on their own target cells. Briefly, androgens also affect the proliferation of their target cells in a two-step sequence. In step 1, androgens promote proliferation by canceling the inhibitory effect of specific serum factors, and in step 2, androgens inhibit the proliferation of these same cells by inducing the synthesis of specific genes coding for intracellular proteins whose role is to prevent cells from entering the cycle [34–36].

## Conclusions

All experimental models are by their very nature 'artificial'. Researchers choose as models, the species or strains, or developmental stages, in which the phenomenon they are interested in is most homogeneous, so that it may be studied without interference from overlapping or unrelated events. For example, on the one hand, in adult ovariectomized rats, a single injection of estrogen results in the proliferation of the epithelial cells of the uterine lining and endometrial glands, while no proliferative activity is seen in the stromal cells. On the other hand, in immature rats, estrogens also induce a proliferative response in the stroma. Estrogens induce ductal development in the mammary gland of rats, while in mice they also induce the development of mammary gland alveoli. Researchers choose a model and trust that all phenomena, even those that appear as contradictory, will be eventually integrated in a coherent fashion.

Organ and primary cultures, along with 'established' cell lines offer a reduction in complexity, together with an increase in 'artificiality'. Researchers compare the relevance of the *in-culture* results to those of their *in-animal* counterparts; these latter are the ones they try to understand. For example, vaginal epithelial cells that do not proliferate *in situ* in immature mice, proliferate maximally in the absence of estrogens in primary cultures [11, 12, 19, 20]. If the researchers facing this phenomenon believe that the default state is *quiescence*, they most likely will dismiss these results as 'artifactual', and move on to explore other models. If, instead, they are persuaded that the default state in all cells is *proliferation*, they will take this result as a perfectly reasonable outcome.

'Established' cell lines (e.g. the human breast MCF7 cells) are widely accepted models for the study of control of cell proliferation regardless of the premises researchers adopt. Proponents of the positive hypothesis interpret the proliferation of these cells in the absence of serum as indicative of their ability to produce their own growth factors, stating that, since they are derived from tumors, they might have found a way to behave 'abnormally' by overcoming the default state of *quiescence* [37]. However, these same MCF7 tumor cells do not proliferate in ovariectomized hosts unless estrogens are administered to the nude mice into which they were inoculated [38]. The positive hypothesis cannot reconcile these *in-culture* and *in-animal* results. Alternatively, the indirect-negative hypothesis acknowledges that when removed from the organism, cells exercise their built-in ability to proliferate. Estrogenless serum inhibits their proliferation *in culture*, and estrogens overcome this inhibition. Only by adopting the premise that *proliferation* is the default state can the pattern of behavior of MCF7 cells be seen to be consistent in the animal and *in culture* [39].

Proponents of the stroma as the obligatory source of control of the proliferation of epithelial cells *via* the production of growth factors demur when considering cell lines as appropriate tools, in spite of the good correlation between data obtained in animals and *in culture*. Others dismiss the results obtained in organ explants, whereby epithelial cells proliferate in the absence of serum, growth factors and estrogens, and regardless of the presence of stroma, as being artifacts resulting from removing the tissues from the organism. How can one, *a priori*, establish a relative scale of artificiality regarding stroma–epithelium recombination, organ culture, primary culture or established cell lines?

Experiments are not done in a theoretical vacuum, and choices about what is relevant and what is not depend on the premises researchers believe are the most likely to represent 'reality'. Admittedly, this is an arbitrary choice, since the default state on the proliferation of cells in metazoa cannot be decided by today's experimental means. The chosen premises also determine the experimental design. This means that often experiments designed from one set of premises cannot be interpreted from the perspective of the opposite ones. For example, only if one suspects that serum may contain an inhibitor of cell proliferation would one test the effect of decreasing serum concentrations to find out when the inhibition was no longer effective. Only a few experiments are interpretable by adopting the opposite premises. For example, the use of serumless medium is predicated by the proponents of positive control, on the assumption that it makes cells quiescent. On the contrary, the proponents of the negative control hypothesis use this artifact to allow cells to express their ability to proliferate. Hence, the fact that estrogen-target cells proliferate in serumless medium is interpreted as meaning that cells secrete their own growth factors by the former, and that they are expressing the default state by the latter.

Our analysis of the available data on the proliferation of sex steroid target cells from the perspective of the negative control hypothesis explains most simply and rigorously the experimental observations. Most importantly, this type of analysis is consistent with the concept that the default state of all cells is *proliferation*, a trait conserved throughout evolution.

## References


1. **Martin, L.** (1980) Estrogens, antiestrogens and the regulation of cell proliferation in the female reproductive tract in vivo. In: *Estrogens in the Environment* (ed. J.A. McLachlan). Elsevier/North-Holland, New York, pp. 103–130.
2. **Stormshak, F., Leake, R., Wertz, N. and Gorski, J.** (1976) Stimulatory and inhibitory effects of estrogen on uterine DNA synthesis. *Endocrinology* **99**: 1501–1511.
3. **Jensen, E.V. and Jacobson, H.I.** (1962) Basic guide to the mechanism of estrogen action. *Recent Prog. Hormone Res.* **18**: 387–406.
4. **Kumar, V., Green, S., Stack, G., Berry, M., Jin, J.R. and Chambon, P.** (1987) Functional domains of the human estrogen receptor. *Cell* **51**: 941–951.
5. **Furth, J., Clifton, K.H., Gadsden, E.L. and Buffet, R.F.** (1956) Dependent and autonomous mammotropic pituitary tumors in rats: their somatotropic features. *Cancer Res.* **16**: 608–616.
6. **Sonnenschein, C. and Soto, A.M.** (1980) But . . . are estrogens per se growth-promoting hormones? *J. Natl Cancer Inst.* **64**: 211–215.
7. **Soto, A.M. and Sonnenschein, C.** (1987) Cell proliferation of estrogen-sensitive cells: the case for negative control. *Endocr. Rev.* **8**: 44–52.
8. **Sirbasku, D.A.** (1978) Estrogen induction of growth factors specific for hormone-responsive mammary, pituitary, and kidney tumor cells. *Proc. Natl Acad. Sci. USA* **75**: 3786–3790.
9. **Sonnenschein, C. and Soto, A.M.** (1978) Pituitary uterotrophic effect in the estrogen-dependent growth of the rat uterus. *J. Steroid Biochem.* **9**: 533–537.
10. **Dickson, R.B. and Lippman, M.E.** (1987) Estrogenic regulation of growth and polypeptide growth factor secretion in human breast carcinoma. *Endocr. Rev.* **8**: 29–43.
11. **Martin, L.** (1959) Growth of the vaginal epithelium of the mouse in tissue culture. *J. Endocrinol.* **18**: 334–342.
12. **Flaxman, B.A., Chopra, D.P. and Harper, R.A.** (1974) Autoradiographic analysis of hormone-independent development of the mouse vaginal epithelium in organ culture. *In Vitro* **10**: 42–50.

13. **Berthois, Y., Katzenellenbogen, J.A. and Katzenellenbogen, B.S.** (1986) Phenol red in tissue culture is a weak estrogen: implications concerning the study of estrogen responsive cells in culture. *Proc. Natl Acad. Sci. USA* **83**: 2496–2500.

14. **Villalobos, M., Olea, N., Brotons, J.A., Olea-Serrano, M.F., Ruiz de Almodovar, J.M. and Pedraza, V.** (1995) The E-screen assay: a comparison of different MCF7 cell stocks. *Environ. Health Perspect.* **103**: 844–850.

15. **Barnes, D. and Sato, G.** (1980) Growth of a human mammary tumor cell line in a serum-free medium. *Nature* **281**: 388–389.

16. **Soto, A.M. and Sonnenschein, C.** (1984) Mechanism of estrogen action on cellular proliferation: evidence for indirect and negative control on cloned breast tumor cells. *Biochem. Biophys. Res. Commun.* **122**: 1097–1103.

17. **Lykkesfeldt, A.E. and Briand, P.** (1986) Indirect mechanism of oestradiol stimulation of cell proliferation of human breast cancer cell lines. *Br. J. Cancer* **53**: 29–35.

18. **Briand, P. and Lykkesfeldt, A.E.** (1986) Long-term cultivation of a human breast cancer cell line, MCF7, in chemically defined medium. *Anticancer Res.* **6**: 85–90.

19. **Uchima, F.D.A., Edery, M., Iguchi, T. and Bern, H.A.** (1991) Growth of mouse endometrial luminal epithelia cells in vitro: functional integrity of the oestrogen receptor system and failure of oestrogen to induce proliferation. *J. Endocrinol.* **128**: 115–120.

20. **Tsai, P.S. and Bern, H.A.** (1991) Estrogen-independent growth of mouse vaginal epithelium in organ culture. *J. Exp. Zool.* **259**: 238–245.

21. **Fukamachi, H. and McLachlan, J.A.** (1991) Proliferation and differentiation of mouse uterine epithelial cells in primary serum-free culture: estradiol-17B suppresses uterine epithelial proliferation cultured on a basement membrane-like substratum. *In Vitro Cell Dev. Biol.* **27A**: 907–913.

22. **Jiang, S-Y. and Jordan, V.C.** (1992) Growth regulation of estrogen-receptor negative breast cancer cells transfected with cDNAs for estrogen receptor. *J. Natl Cancer Inst.* **84**: 580–591.

23. **Laursen, I., Briand, P. and Lykkesfeldt, A.E.** (1990) Serum albumin as a modulator on growth of the human breast cancer cell line, MCF-7. *Anticancer Res.* **10**: 343–351.

24. **Sonnenschein, C., Soto, A.M. and Michaelson, C.L.** (1996) Human serum

albumin shares the properties of estrocolyone-I, the inhibitor of the proliferation of estrogen-target cells. *J. Steroid Biochem. Mol. Biol.* **59**: 147–154.

25. **Huet-Hudson, Y.M., Chakraborty, C., De, S.K., Suzuki, Y., Andrews, G.K. and Dey, S.K.** (1990) Estrogen regulates the synthesis of epidermal growth factor in mouse uterine epithelial cells. *Mol. Endocrinol.* **4**: 510–523.

26. **Stancel, G.M., Chiappetta, C., Gardner, R.M., Kirkland, J.L., Lin, T.H., Lingham, R.B., Loose-Mitchell, D.S., Mukku, V.R. and Orengo, C.A.** (1990) Regulation of the uterine epidermal growth factor receptor by estrogen. *Prog. Clin. Biol. Res.* **322**: 213–226.

27. **Nelson, K.G., Takahashi, T., Bossert, N.L., Walmer, D.K. and McLachlan, J.A.** (1991) Epidermal growth factor replaces estrogen in the stimulation of female genital-tract growth and differentiation. *Proc. Natl Acad. Sci. USA* **88**: 21–25.

28. **Curtis, S.W., Washburn, T., Sewall, C., Diaugustine, R., Lindzey, J., Couse, J.F. and Korach, K.S.** (1996) Physiological coupling of growth factor and steroid receptor signaling pathways: estrogen receptor knockout mice lack estrogen-like response to epidermal growth factor. *Proc. Natl Acad. Sci. USA* **93**: 12626–12630.

29. **Ignar-Trowbridge, D.M., Nelson, K.G., Bidwell, M.C., Curtis, S.W., Washburn, T.F., McLachlan, J.A. and Korach, K.S.** (1992) Coupling of dual signaling pathways: epidermal growth factor action involves the estrogen receptor. *Proc. Natl Acad. Sci. USA* **89**: 4658–4662.

30. **Cunha, G.R. and Young, P.** (1996) EGF receptor (EGFR) signaling plays different roles in estrogen-induced epithelial proliferation in mammary gland (MG) versus uterus (UT). *Int. Congress Endo.* **2**: 740 (Abstract).

31. **Cunha, G.R., Bigsby, R.M., Cooke, P.S. and Sugimura, Y.** (1985) Stromal–epithelial interactions in adult organs. *Cell Differentiation* **17**: 137–148.

32. **Hayward, S.W., Rosen, M.A. and Cunha, G.R.** (1997) Stromal–epithelial interactions in the normal and neoplastic prostate. *Br. J. Urol.* **79**(Suppl 2): 18–26.

33. **Bruchovsky, N., Lesser, B., Van Doorn, E. and Craven, S.** (1975) Hormonal effects on cell proliferation in rat prostate. *Vit. & Horm.* **33**: 61–102.

34. **Sonnenschein, C., Olea, N., Pasanen, M.E. and Soto, A.M.** (1989) Negative controls of cell proliferation: human prostate cancer cells and androgens. *Cancer Res.* **49**: 3474–3481.

35. **Soto, A.M., Lin, T.M., Sakabe, K., Olea, N., Damassa, D.A. and Sonnenschein, C.** (1995) Variants of the human prostate LNCaP cell line as a tool to study discrete components of the androgen-mediated proliferative response. *Oncology Res.* **7**: 545–558.

36. **Geck, P., Szelei, J., Jimenez, J., Lin, T.M., Sonnenschein, C. and Soto, A.M.** (1997) Expression of novel genes linked to the androgen-induced, proliferative shutoff in prostate cancer cells. *J. Steroid Biochem. Mol. Biol.* **63**: 211–218.

37. **Lippman, M.E., Dickson, R.B., Gelmann, E.P., Rosen, N., Knabbe, C., Bates, S., Huff, K. and Kasid, A.** (1987) Growth regulation of human breast carcinoma occurs through regulated growth factor secretion. *J. Cell. Biochem.* **35**: 1–16.

38. **Soule, H.D. and McGrath, C.M.** (1980) Estrogen responsive proliferation of clonal human breast carcinoma cells in athymic mice. *Cancer Lett.* **10**: 177–189.

39. **Soto, A.M. and Sonnenschein, C.** (1985) The role of estrogens on the proliferation of human breast tumor cells (MCF-7). *J. Steroid Biochem.* **23**: 87–94.

## Chapter 6

# Cell Proliferation and Tissue Differentiation

"The fundamental problem of chemical physiology and of embryology is to understand why tissue cells do not all express, all the time, all the potentialities inherent in their genome [that is, in the totality of their genes]. The survival of the organism requires that many, and in some cases most, of these potentialities be unexpressed, that is to say *repressed*."

*Jacob, F. and Monod, J. (1961) Genetic regulatory mechanisms in the synthesis of proteins. J. Mol. Biol.* **3**: *318–356*

". . . Philosophical difficulties . . . have rarely stood as impediments of scientific discourse. Philosophical difficulties are traditionally shunted aside until the moment in history when the contradiction in question renders it impossible to incorporate some particularly glaring empirical data into the existing framework . . ."

*Leo W. Buss (1987) The Evolution of Individuality. Princeton University Press, Princeton, NJ, p. 177, footnote 9*

## The different meanings of 'differentiation'

*All* cells in multicellular organisms are 'differentiated'. However, it is frequently heard in scientific seminars and lectures, or read in scientific and lay publications, that differentiated cells do not proliferate, and/or that cells that proliferate are undifferentiated [1]. This tautology or circular argument should be routed out from the biological discourse. These statements are misleading and affect not only biological thinking but, as we will see in Part II, also our understanding of cancer.

The definition of 'differentiation' has been the subject of much controversy because it is inherently linked to comparisons between and among shifting phenotypes. Differentiation deals with *how* the diverse, 'specialized' cell types in an organism are derived from a single cell. In this regard, differentiation may be considered as a relative and qualitative concept. By qualitative, we mean that a cell is equally differentiated when synthesizing keratin, as epithelial cells do, as another cell that synthesizes acetylcholine, as neurons do. Confusion sets in when differentiation is used as a quantitative concept. For example, it has been used to mean that one cell type, say a fibroblast, is morphologically and functionally less differentiated than say, a nerve cell. In fact, neurons, fibroblasts and keratinocytes express different sets of genes, rather than any of them being any more 'differentiated' than the others.

Differentiation also serves to define the process of self-renewal by *stem cells* in adult tissues. In this case, differentiation becomes a quantitative and hierarchical process; the example is set by a maturation sequence whereby stem cells, unable to perform the function of a particular tissue, generate cells that progressively differentiate into a fully functional cell. Hemopoiesis is one example of this situation that will be dealt with later in this chapter. The ensuing confusion on the subject may represent no more than a nuisance for those of us who have been exposed to this almost irrational tradition for a good part of our scientific careers; more damaging, perhaps, is the confusion generated in young students who are approaching the already monumental complexity of biological subjects.

## A hierarchical perspective

Biology is extremely complex. To simplify the acquisition of complex knowledge, textbooks, professors and teachers introduce some simple-minded notions so that students are not overwhelmed. Confusion often results when complexity is explained 'in a nutshell' as often requested by the lay person, the science writer, or even other scientists working in other disciplines.

A full-size, adult, multicellular organism is produced out of a single cell. Self-renewal in some tissues occurs by simple duplication of their differentiated cells, while in other tissues, it is accomplished by *stem cells*. There are several types of stem cells. Those which retain the widest capabilities are considered *totipotential*; the ultimate totipotential cell is the fertilized egg. *Pluripotential* stem cells may generate cells of several distinct lineages. Among the best studied pluripotent stem cells are the hemopoietic ones that generate blood cell types and the cells in the crypts of the intestinal mucosa. Daughters of pluripotent stem cells may become restricted in their potentiality, and hence, generate only one cell type. The epithelial germinative cells of the epidermis are considered *monopotent* stem cells.

Four decades ago, it was shown that a single carrot cell could generate a whole carrot plant [2]. Also in the 1950s and 60s, the nuclei of frog intestinal cells transplanted into oocyte cytoplasms generated whole individuals [3, 4]. When hierarchical barriers vanish, cells, or their nuclei, may become indeed *pluri* or *totipotential*. Dolly, the Scottish parent-less sheep generated from the nucleus of an adult mammary gland cell transplanted into an enucleated sheep oocyte cytoplasm is the latest famous example of the totipotentiality of the genome.

*In culture*, cells from metazoa show properties that they do not demonstrate at the organismal, organ, or tissue levels. By virtue of becoming freed from the constraints of the homeostatic influences necessary to coordinate the needs of a multicellular organism, the 'liberated' metazoan cell *in culture* may re-acquire ancestral, cryptic properties, including proliferation and mobility. 'Demergence' has taken place.

## The elusive concept of stem cells

Stem cells are cells with a high capacity for self-renewal. However, no morphologic or biochemical sign unequivocally characterizes a stem cell. In a histological slide, one may infer which may be a stem cell by its position. When disrupted from this hierarchical tissue context, stem cells are difficult to distinguish. Therefore, stem cells are defined by the assay used to detect them, that is, this is essentially an operational definition. In view of these uncertainties, it is entirely apt to compare the concept of stem cells to that of pornography for US Supreme Court justice Byron White who was quoted as saying 'It's hard to define it, but I know when I see it' [5].

As mentioned above, *all* cells in multicellular organisms are differentiated; this statement applies of course to stem cells. This is not a **semantic** issue. That undifferentiated stem cells generate differentiated daughter cells is a relativistic notion. All stem cells must have receptors for cues that tightly control their own proliferation. When they proliferate, they are sufficiently differentiated as to direct one of their daughter cells to either acquire properties other than those of their progenitor (A → A + B, where A means stem cell, and B means differentiated cell), or to generate two indistinguishable daughter stem cells (A → A + A). In some tissues, the 'asymmetric division' (A → A + B) obeys a positional strategy: one daughter cell remains where its parent was, while the other moves along to other niches that will favor the expression of other phenotypes. Proliferation may or may not be part of the next stage in the developmental process. On average, stem cells lodged in the deepest, basal layer of the skin epithelium in vertebrates generate one stem cell and one migrating cell (A → A + B). In this example, B cells will not proliferate anymore; instead, they will change their gene expression repertory and will eventually die, or be sloughed off. From our perspective, both A and B cells are differentiated, each having very specific, unique tasks. Incidentally, by abrading the epidermis and eliminating cells in the superficial layers, a wave of mitosis is elicited in the deepest germinative cell layer. This suggests that, while intact, the former send inhibitory (paracrine) cues toward the cells in the germinative layer; a reduced inhibition would trigger the initiation of the proliferative event. This is another example of how switching paradigms to one where *proliferation* is the default state makes explanations more parsimonious.

## The hemopoietic stem cells

In vertebrates, red and white blood cells are generated continuously throughout life in specialized organs, like the thymus, the liver, the bone marrow and the spleen, or their equivalents in different vertebrates and invertebrates. In their midst, under normal conditions, hemopoietic cells generate staggering numbers of blood cells. In humans, for example, 2.5 billion red cells, 1 billion granulocytes and 1.5 billion platelets per kilogram of body weight are generated daily [6].

### *Stem cells in the bone marrow of vertebrates*

Stem cells are operationally defined by the assay system used to detect them [7, 8]. A few weeks after marrow cells are injected into lethally irradiated mice, 'high frequency' pluripotent cells contribute to the repopulation of hemopoietic organs, while months later, few pluripotent cells are active. This led Müller-Sieburg and Deryugina to propose that stem cells could be further classified operationally into 'root' cells, which are the most pluripotent stem cells having long-term repopulation capability, 'trunk' cells, which are responsible for rapid onset repopulation, but have limited persistence, and 'branch' cells, which are lineage-restricted [7].

In the bone marrow, as in any other tissue, the position that a cell occupies has informational content. After each cell division, daughter cells migrate to progressively greater distance from the stem cell environment; they then occupy niches that promote lineage-commitment and maturation [9]. A scenario postulated over 30 years ago by Wolf and Trentin is now being explored extensively: in hematopoietic organs, stromal cells are located adjacent to stem cells [10]. Müller-Sieburg and Deryugina proposed that 'a major role of the stroma lays in protecting stem cells from proliferation and thereby retaining the stem cell pool'. This can easily be translated into the negative control hypothesis vocabulary as follows: stem cells have a dominant, constitutive capacity to proliferate. Stroma cells would inhibit stem cell proliferation by means of discrete molecules secreted tonically into their immediate surrounding, or cell-to-cell interactions through junctional complexes. In addition, the evidence is also compatible with the notion that hemopoietic stromal cells provide nutritional components that prevent stem cells from entering apoptosis, therefore, providing for the survival of their adjacent stem cells. Organismal 'distress signals' such as low oxygen pressure, anemia, *et cetera,* may be read by the stromal cells as an order to cease imparting inhibitory signals to the stem cells that are in their close proximity. This interpretation does not require the intervention of stimulators of cell proliferation.

### *The hormone EPO: growth factor or survival factor?*

EPO is a hormone produced by the kidney; its plasma level is controlled by the oxygen pressure in the outside environment. Inhabitants of the Andes, and the Himalayan mountains, have a higher concentration of erythrocytes (polycythemia) than individuals living at sea level. When the latter move (even temporarily) to higher elevations, their cardiac and intellectual performance quickly reflect the lower oxygen pressure. EPO circulates at higher concentrations in highlanders than in lowlanders; with time, it increases in 'transplanted' lowlanders and, subsequently, their red blood cell concentrations also increase. Thanks to recombinant DNA technology, purified EPO is now available to be used experimentally and for therapy. EPO has also been used illegally by athletes who wish to improve their performance.

For the purpose of this discussion, it is important to analyze the data and their interpretations about how EPO increases the number of erythrocytes. At first, it was thought that EPO directly stimulated the proliferation of erythroid stem cells. However, after extensive research on the subject, supporting evidence for a stimulatory role of EPO is lacking. In fact, no EPO-target cells can be maintained alive *in culture* in a resting, prolonged $G_0/G_1$ state in its absence. Instead, these cells **die** as a result of EPO deprivation [11]. In addition, proliferation of erythroid precursor cells in bone marrow still occurs in animals deprived of EPO; this suggests that EPO regulates erythropoiesis by controlling the level of cell death of erythroid precursor cells [12]. Remarkably, EPO passes the knockout 'acid test': the lack of EPO or its receptor result in fetal death due to impaired erythropoiesis [13].

Other factors implicated in erythropoiesis such as the stem cell factor (SCF), also called kit ligand, and Insulin-like Growth Factor-I have been shown to inhibit apoptosis *in culture* at levels that support colony formation. In fact, the most parsimonious interpretation of these data is that hemopoietic growth factors inhibit cell death, and allow the 'differentiation' program to be expressed unhindered.

### *The maturation of blood cells*

In the last two decades, much work has been done to identify signals believed to be responsible for the maturation of hemopoietic stem cells. Cell culture models have been used for this purpose. These experimental models relied on the premise that cells are stuck in the quiescent state; implicitly, this rationale acknowledges that putative growth factors would stimulate cells to enter the cycle. The formation of maturing granulocyte and macrophage colonies in semisolid media required co-culture with cells from lung, heart, spleen and uterus (of these, only the spleen is a hemopoietic organ), or with medium conditioned by these cells. Fractionation of these conditioned media led to the purification of glycoproteins that were operationally called *colony-stimulating factors* (CSFs) [14]. Others were more cautious and gave these putative effectors a generic, less compromising name, *cytokines*. Still others prefer to call them *restrictins* under the assumption that they affect their target cells by narrowing their options from the wider repertory held by their parental cells [15].

What is meant exactly by 'stimulation' in the context of clonal assays? The clonal assays in semisolid culture systems exploit the ability of these stem cells to form clones of maturing blood cells. These assays, therefore, measure the total number of cells obtained from a single stem cell, the number of morphologically identifiable cells at different stages of maturation, and whether the maturing cells belong to one or more lineages. This assay does not reveal, however, whether this increase in cell number is due to increased survival, inhibition of 'programmed cell death' (apoptosis), or stimulation of cell proliferation; it should be remembered that this assay has been designed under the tacitly accepted premise that *quiescence* is the default state in metazoa. Hence, it was taken for granted that these CSFs were signals that stimulated quiescent cells to proliferate (growth factors). However, even the staunchest proponents of a proliferative role for these factors acknowledge that they may act to enhance, instead, the *viability* of precursors and mature cells [14]. Later on, these factors were found to inhibit apoptosis of erythroid and of granulocyte progenitor cells [12].

In conclusion, first, EPO and the CSFs have not been proven to be stimulators of cell proliferation; and second, data are compatible, instead, with the notion that they are inhibitors of cell death. In addition, data generated using cytokine knockouts are now challenging the notion that they play a significant role in the proliferation of their putative target cells [16]. The reassessment made above also highlights the need to use *in animal–in culture* models in order to pose causal questions at the pertinent hierarchical level within a physiological context.

## Proliferation of antibody-producing cells: a different perspective

The immune system in vertebrates plays a vital role in defense mechanisms. The immune response is a multifaceted, complex process. Here, we will only address the rationale used to explain the proliferative events that take place during the immune response.

Several cell types contribute to the immunological functions. Among them, B- and T-lymphocytes are at the center of the response to foreign antigens, be they soluble or particulate (chemicals, viruses, bacteria, etc.). The former synthesize and secrete antibodies, and are produced in the bone marrow. The latter deal with cell-mediated immune responses (e.g. to kill infected cells, or reject foreign tissue) and are produced in the thymus. Both types of lymphocytes have unique cell surface specific receptors that allow them to recognize particular antigens that they have never seen before.

Burnett proposed the *clonal selection theory* to explain the diversity and specificity of the immune responses [17]. The diversity of antibodies has been explained by the rearrangement of the antibody genes [18]. Once lymphocytes mature in the primary lymphatic organs (bone marrow and thymus), they enter the general circulation; these are called 'virgin' or 'naive' cells. The cells that come in contact with their specific antigens, are 'activated', proliferate, and begin to synthesize and secrete antibodies. Other virgin cells exposed to the same specific antigens also proliferate; however, they do not secrete antibodies. They are kept in a pool of cells called 'memory' cells; these become activated in subsequent encounters with the same antigen and also undergo clonal expansion.

Given that we have proposed a sweeping reassessment of the premises used to interpret data on the control of cell proliferation in metazoa, B- and T-lymphocytes cannot be ignored within this generalization. As with other cells considered in this book, we pose the questions: Is the default state of virgin and memory cells, *proliferation* or *quiescence*? Is it reasonable to propose that virgin and memory lymphocytes are not just 'unstimulated' cells, but are actively inhibited from proliferating? Moreover, are the data on 'activation' of B- and T-lymphocytes interpretable from the perspective of the negative control hypothesis? Also, more explicitly, can antigens, antigen-presenting cells and cytokines, alone or in combination, be neutralizers of inhibitors of the proliferation of B- and T-lymphocytes?

### *Are antigens growth factors, or neutralizers of inhibitory factors?*

Virgin B- and T-lymphocytes are the last link in a lineage of white blood stem cells. These lymphocytes are thought to be 'terminally differentiated' cells. However, these arrested cells can be quickly switched into rapidly proliferating cells when exposed to specific antigens; this reaction triggers their clonal expansion. In the case of T-lymphocytes, the antigen (in the form of an antigen-presenting cell) and second signals (B7 and CD28) are responsible for clonal expansion and differentiation [19]. The proliferation and subsequent reversal of this expansion are subjects of interest for their biological and medical implications.

An explanation for the proliferative event differs depending on which premise is adopted in this regard. By considering *quiescence* as the default state, antigens become growth factors; instead, by adopting *proliferation* as the default state, the antigen becomes the means to cancel the inhibition which the virgin or memory B- and T-lymphocytes are under. This does not represent a semantic option when taking into account the historical record of these cells. These lymphocytes, as all the other cells in the body of animals, originate from a fertilized egg, whose own default state was *proliferation.* In Chapter 2, we argued for the maintenance of *proliferation* as the default state of *all* cells, including, of course, those in metazoa. If cells in these multicellular organisms appear as being in a quiescent state, this is likely to be due to an active induction mechanism; examples of this contingency were mentioned in Chapter 5 when referring to the proliferative shutoff effects triggered by sex hormones on their respective target cells (estrogens on the uterus, mammary gland and hypophysis, and androgens on the prostate gland).

If these lymphocytes are also subject to the regularity of conserving *proliferation* as the default state of all cells, one may propose that these cells are caught in a temporary, switchable proliferative shutoff. Under this revised scenario, the presence of B-cell receptors in their plasma membranes would represent the negative signal for proliferation. B-lymphocytes would be recruited into a proliferative pool of cells by coming in contact with a unique antigen that, by recognizing the B-cell receptor, would unlock the built-in capacity of these cells to proliferate.

It has been proposed that, in addition to antigens, the proliferation of B- and T-lymphocytes is stimulated by complex interactions in which helper T-lymphocytes play a significant role. It should be mentioned that most of the data on this subject have been collected using *in-culture* models. Several cytokines have been proposed to have a stimulatory effect on the proliferation of B- and T-lymphocytes when they become engaged in clonal expansion [20, 21]; however, these cytokines appear to be neither necessary nor sufficient to trigger the proliferation of these cells in animal models. In the case of T-lymphocytes, interleukin-2 or IL-2 ('T-cell growth factor') was claimed to be essential for their proliferation *in culture*. Remarkably, however, a requirement for the use of IL-2 is not absolute even in culture conditions [22]. Moreover, the role of this putative growth factor has not been validated using cytokine knockouts. For instance, when the IL-2 gene was disabled in mice, or was missing in a human patient, the cell populations in the thymus were unaffected, and the composition of the circulating T-cell subsets was not compromised [23, 24]. Even when double knockouts (IL-2 plus IL-4) were tested, the proliferation of T-cells was not affected; this weakens the argument that there is redundancy in the control of the proliferation of these cells [16, 25]. In regard to B-lymphocytes, the cytokines IL-4 and IL-6 have also been claimed to be crucial for their proliferation *in culture*. However, the evidence provided by cytokine knockouts has not been supportive of this notion [26, 27]. Taken together, the putative proliferative activity of cytokines *in culture* has not received support from the use of knockout mice, clearly a worthy experimental tool aimed at verifying the suspected physiological roles of growth factors [28].

In summary, it is entirely plausible to assign a role to these cytokines as survival factors *in culture*, or as 'differentiation' factors mainly engaged in the maturation of lymphocytes along the complex path they follow to react finally with infectious microorganisms or 'foreign tissues' in the animal. The notion that *proliferation* is the default state in all cells dispenses with the need to search for elusive positive signals that would actively induce these lymphopoietic stem and precursor cells to enter the cycle. This conclusion is based on the data collected by knocking-out genes whose products have been linked to putative stimulatory signals. A direct proliferative role played by those cytokines should probably be revised based on the arguments raised in this book. While the positive and the negative hypotheses are operationally equivalent, the notion that *proliferation* is the default state of all cells is consistent with ontogenic and phylogenic arguments (see Chapters 1 and 2).

## Conclusions

The sweeping statement '*All* cells in multicellular organisms are differentiated' was aimed at calling attention to the fact that each cell type in a multicellular organism is there because it plays a specific role in that living individual. By approaching biology from a hierarchical perspective, it should be plain that a brain cell is not more important than a heart muscle cell; they belong to a similar hierarchical level, and they are equally necessary. They do not compete between themselves for a position of preeminence among equals.

Several conclusions may be drawn when published data on proliferation and differentiation are put in the context of the default state being either *proliferation* or *quiescence*. They are:

- The statement '. . . differentiated cells do not proliferate; only undifferentiated cells do' is contradicted by evidence. Tissue self-renewal may occur as a result of the proliferation of functional differentiated cells in some organs, and of stem cells in others. Hence, there is no univocal relation between proliferation and differentiation. Cell proliferation and differentiation are biological phenomena that are best analyzed by dissociating them conceptually.
- The interpretation of data on the control of cell proliferation will depend on the premise the researcher adopts to study the subject. Hemopoiesis is a highly regulated process involving endocrine, paracrine and cell-to-cell control mechanisms. Stem cells proliferate at a slow rate and may be controlled negatively by neighboring stroma. Blood cell numbers are regulated by multifunctional factors; EPO, for instance, enhances cell numbers by inhibiting the death of maturing cells. Our analysis of the evidence favors the notion that cytokines may be survival and maturation factors.
- Gene targeting experiments (knockouts) have administered 'the acid test' to the perceived role of growth factors and cytokines. These latter concepts stem from experiments done *in culture* under the premise that *quiescence* is the default state in metazoa. The evidence from knockout experiments contradicts the conclusions that growth factors and cytokines are stimulators of cell proliferation.
- All along this book, we have been making the case that *proliferation* is the default state of all cells. B- and T-lymphocytes are perhaps the most difficult examples to integrate within the perspective we favor. This can be dealt with by either claiming that lymphocytes are a special case among all cells, where *quiescence* is the default state. An alternative option would propose to conceptually switch to the notion that antigens, in their different forms, are agents that specifically release (inhibit the inhibition) the built-in capacity of these lymphocytes to proliferate. For the sake of biological relevance, we favor the latter option since ontogenetically lymphocytes originated from cells (beginning with the egg) that should have had *proliferation* as their default state.

These incongruences between the positive hypothesis for the control of cell proliferation and the data gathered by exploring it suggest that a new strategy should be considered. The alternative working hypothesis based on the premise that *proliferation* is the default state for all cells offers the opportunity to go back to the drawing board and develop such a new strategy.

## References

1. **Darnell, J., Lodish, H. and Baltimore, D.** (1986) *Molecular Cell Biology.* Scientific American Books, New York, pp. 1030–1031.

2. **Steward, F.C., Mapes, M.O. and Smith, J.** (1958) Growth and organized development of cultured cells. I. Growth and division freely suspended cells. *Am. J. Botany* **45**: 693–703.

3. **Briggs, R. and King, T.J.** (1952) Transplantation of living nuclei from blastula cells into enucleated frogs' eggs. *Proc. Natl Acad. Sci. USA* **38**: 455–463.

4. **DiBerardino, M.A., Orr, N.H. and McKinnell, R.G.** (1986) Feeding tadpoles cloned from Rana erythrocyte nuclei. *Proc. Natl Acad. Sci. USA* **83**: 8231–8234.

5. **Morrison, S.J., Shah, N.M. and Anderson, D.J.** (1997) Regulatory mechanisms in stem cell biology. *Cell* **88**: 287–298.

6. **Erslev, A.J. and Lichtman, M.A.** (1990) Structure and function of the marrow. In: *Hematology* (eds W.J. Williams, E. Beutler and A.J. Erslev). McGraw-Hill, New York, pp. 37–47.

7. **Muller-Sieburg, C.E. and Deryugina, E.** (1995) The stromal cells' guide to the stem cell universe. *Stem Cells* **13**: 477–486.

8. **Weisman, I.L.** (1994) Developmental switches in the immune system. *Cell* **76**: 207–218.

9. **Uchida, N., Fleming, W.H., Alpern, E.J. and Weissman, I.L.** (1993) Heterogeneity of hematopoetic stem cells. *Curr. Opinions Immunol.* **5**: 177–184.

10. **Wolf, N.S. and Trentin, J.J.** (1968) Hemopoietic colony studies V. Effects of hemopoietic organ stroma on differentiation of pluripotent stem cells. *J. Exp. Med.* **12**: 205–214.

11. **Koury, M.J. and Bondurant, M.C.** (1988) Maintenance by erythropoietin of viability and maturation of murine erythroid precursor cells. *J. Cell Physiol.* **137**: 65–74.

12. **Williams, G.T., Smith, C.A., Spooncer, E., Dexter, T.M. and Taylor, D.R.** (1990) Haemopoietic colony stimulating factor promotes cell survival by suppressing apoptosis. *Nature* **343**: 76–79.

13. **Wu, H., Liu, X., Jaenisch, R. and Lodish, H.F.** (1995) Generation of committed erythroid BFU-E and CFU-E progenitors does not require erythropoietin or the erythropoietin receptor. *Cell* **83**: 59–67.

14. **Metcalf, D.** (1989) The molecular control of cell division, differentiation commitment and maturation in hematopoietic cells. *Nature* **339**: 27–30.

15. **Zipori, D.** (1990) Regulation of hemopoiesis by cytokines that restrict options for growth and differentiation. *Cancer Cells* **2**: 205–211.

16. **Durum, S.K. and Muegge, K.** (1998) Preface. In: *Cytokine Knockouts* (eds S.K. Durum and K. Muegge). Humana Press, Totowa, NJ, pp. vii–xvi.

17. **Burnett, F.M.** (1959) *The Clonal Selection Theory of Immunity.* Vanderbilt University Press, Nashville, TN.

18. **Hozumi, N. and Tonegawa, S.** (1976) Evidence for somatic rearrangement of immunoglobulin genes coding for variable and constant regions. *Proc. Natl Acad. Sci. USA* **73**: 3628–3632.

19. **Parijs, L.V. and Abbas, A.K.** (1998) Homeostasis and self-tolerance in the immune system: turning lymphocytes off. *Science* **280**: 243–248.

20. **Fitch, F.W., Lancki, D.W. and Gajewski, T.F.** (1993) T-cell mediated immune regulation. In: *Fundamental Immunology* (ed. W.E. Paul). Raven Press, New York, p. 733.

21. **Howard, M.C., Miyajima, A. and Coffman, R.** (1993) T-cell-derived cytokines and their receptors. In: *Fundamental Immunology* (ed. W.E. Paul). Raven Press, New York, p. 763.

22. **Laing, T.J.** (1988) IL-2 independent proliferation in human T cells. *J. Immunol.* **140**: 1056–1062.

23. **Schorle, H., Holtschke, T., Hunig, T., Schimple, A. and Horak, I.** (1991) Development and function of T cells in mice rendered interleukin-2 deficient by gene targeting. *Nature* **352**: 621–624.

24. **Weinberg, K. and Parkman, R.** (1990) Severe combined immunodeficiency due to a specific defect in the production of interleukin-2. *N. Engl. J. Med.* **322**: 1718–1723.

25. **Sadlack, B., Kuhn, R., Schorle, H., Rajewsky, K., Muller, W. and Horak, I.** (1994) Development and proliferation of lymphocytes in mice deficient for both interleukin-2 and -4. *Eur. J. Immunol.* **24**: 281–284.

26. **von Freeden, J.U., Moore, T.A., Zlotnik, A. and Murray, R.** (1998) IL-7 knockout mice and the generation of lymphocytes. In: *Cytokine Knockouts* (eds S.K. Durum and K. Muegge). Humana Press, Totowa, NJ, pp. 21–36.

27. **Ramsay, H.A. and Kopf, M.** (1998) IL-6 gene knockout mice. In: *Cytokine Knockouts* (eds S.K. Durum and K. Muegge). Humana Press, Totowa, NJ, pp. 227–236.

28. **Durum, S.K. and Muegge, K.** (1998) *Cytokine Knockouts*. Humana Press, Totowa, NJ, pp. VII–XVI.

## Chapter 7

# Introduction to Carcinogenesis and Neoplasia

"At the time of this writing, the somatic mutation hypothesis, after more than half a century, remains an analogy: 'it is presumptive reasoning based on the assumption that if things have similar attributes they will have other similar attributes' (The Shorter Oxford English Dictionary, Clarendon Press, Oxford, 1947). No discoveries that tighten this analogy have been made. Of course, one can assume that tumors are due to somatic mutations of some qualitative type unrecognized as yet, thus piling surmise on surmise."

*Peyton Rous (1959) Surmise and fact on the nature of cancer. Nature* **183**: *1357–1361*

"We can imagine a time coming when the nature of (DNA) sequence changes in each cancer is used to determine the likely cause."

*John Cairns (1997) Matters of Life and Death. Princeton University Press, Princeton, NJ, p. 164*

Literally, neoplasia means new growth. After more than a century of research, however, definitions of neoplasia are plagued with problems that stem from our imperfect grasp of the biological process that underlies its genesis. Historically, the study of neoplasia interested physicians, who were concerned with curing patients with 'tumors' (etymologically, lumps). Physicians described the disease, and divided 'tumors' into two categories, benign and malignant, according to their clinical behavior. In general, benign tumors are circumscribed, often separated from the adjacent normal tissues by a capsule; they grow slowly, and for the most part, they are not life-threatening. Malignant tumors, instead, grow invasively into adjacent tissues, and often give rise to secondary tumors (metastases), that appear in distant organs. After treatment, malignant tumors frequently recur either locally or at a distance from the primary location. Many patients die as a result of these neoplasias. These generalizations, however, hide a good number of exceptions.

In the late 19th century, pathologists began describing the histological pattern of tumors using the light microscope. The means to diagnose cancer have not changed that much since then. In this regard, it is ironic to read in a leading pathology textbook published in 1967 ' . . . It is quite remarkable that in a day when man is on the threshold of mastering outer space, the diagnosis of cancer still rests principally on the subjective impression of the pathologist . . . ' [1]. However, the expectation by pathologists that a type of objective certainty would be ushered in by the molecular biology revolution has not materialized. The light microscope remains the tool with which pathologists diagnose cancer. This simple realization stealthily suggests that tissue disorganization is at the core of carcinogenesis and neoplasia.

For the most part, neoplasias retain the distinctive structures that characterize the organ of origin. The nomenclature of neoplasms used by pathologists and other specialists indicates the tissue source from which they derive. A benign neoplasia of epithelial origin is called 'adenoma' (from gland-like structures) or 'papilloma' (wart-like structures). A malignant neoplasia of epithelial origin is called a 'carcinoma'. Benign neoplasias of connective tissues are called 'fibroma' (from fibrous tissue), 'osteoma' (from bone), and so on. Their malignant counterparts are called 'fibrosarcoma' and 'osteosarcoma', respectively. This nomenclature is neutral on etiological meaning.

Histologic studies reveal a continuum of variation among neoplasias originating in a particular organ. Pathologists introduced the concept of 'differentiation' to denote the degree of deviation found between the architecture of the neoplasia and that of the normal tissue from which it presumably arose. The closer to the normal tissue, the more 'differentiated' the tumor is considered, and conversely, tumors that are less similar to the tissue of origin are called 'anaplastic' or 'undifferentiated'. It was also recognized that, like normal organs, neoplasias also contained a *parenchyma* (the distinctive cell type of an organ) and a supporting tissue or *stroma* (representing the scaffolding to which the parenchymal cells are attached). As discussed in Chapter 6, for their normal development and function, tissues require a normal architecture where parenchymal and stromal cells operate in a complementary fashion. Both cell types are necessary for the normal function by the tissues and organs. Pathologists also observed that metastases often reproduced the structure of the primary tumor. This suggested that neoplastic cells that migrated through blood and lymph, carried, in themselves, the program to recast the tissue of origin.

How do neoplasias arise? The observation of Sir Percival Pott in the 18th century, that chimney sweeps developed scrotal cancer, hinted that chemicals present in coal tar were acting as carcinogens. Environmental agents have taken a more prominent etiological role in the current century due to the indiscriminate introduction of new synthetic chemicals. The discovery of microbes as causes of contagious diseases served as models for the transplantation of neoplasias. The finding that 'filtrable factors' (viruses) generated sarcomas in domestic fowl, suggested a viral etiology for other tumors. All these phenomena stimulated the experimental study of neoplasia, and the development of animal models. 'Spontaneous' tumors were observed in many multicellular species; this showed that neoplastic development was possible in all metazoa.

Clinical pathology and tumor transplantation helped to partially clarify matters. In its beginning, tumor transplantation was fraught with epistemologic problems [2]. At the end of the last century and the beginning of the current one, the subject of graft rejection, and the role of the immune system in this process, had just begun to be addressed [2]. Nonetheless, some tumors were successfully transplanted among individuals of the same species, while transplantation into heterologous species, by and large, always failed. A better understanding of graft rejection stimulated the development of inbred strains of animals. Inbred animals are derived from brother–sister matings; after dozens of crossings, siblings become genetically homogeneous (these are called syngeneic animals). A tumor can then be transplanted from the animal in which it originated to other members of the same inbred strain, and from one transplanted animal to another, thus outliving the organism of origin.

## If normal cells beget normal cells, and neoplastic cells beget neoplastic cells, what causes normal cells to become neoplastic?

Transplantation allowed for the study of the behavior of advanced neoplastic stages; it did not and could not, however, answer questions about the genesis of these neoplasias (carcinogenesis). Transplantation studies revealed that only a fraction of the transplanted tumor cells survived in the host and that these cells were from the parenchyma. The transplanted stroma was lost. Cells from the host make up the 'new' stroma of the transplanted tumor. The finding that the parenchymal cell lineage was propagated and bred true throughout many serial transplants, and the observation that, in spontaneous neoplasia, metastases usually displayed an architecture similar to that of the primary tumor, suggested that the parenchymal cells carried in themselves all the information necessary to form a tumor. This concept led to the question, 'If neoplastic cells breed true to type, and their normal counterparts do so as well, what kind of change occurred in the normal cells that made them neoplastic?' Enter the notion of mutation. Mutation is derived from the latin word for change. Hence, it was proposed that *a* neoplastic cell resulted from *a* 'mutated' normal cell. The usage of the word mutation has changed since then; today it means a change in the linear structure of DNA. The *somatic mutation theory*, the currently prevalent theory of carcinogenesis, is based on this assumption.

But . . . is it necessary to invoke genomic (DNA) mutations to explain the different behavior of normal and neoplastic cells? All cells in the adult human organism contain the same information. The DNA in a liver cell is identical to the DNA in a kidney cell of the same individual; however, liver cells beget liver cells while kidney cells beget kidney cells. Moreover, we know that all the information necessary to generate a whole animal is present in the DNA of each somatic cell. The most recent vindication of this fascinating concept was provided by Dolly, the famous Scottish sheep whose genome was once in a cell of the mammary gland of another sheep. Hence, a change in the appearance and/or behavior of a cell (phenotype) does not require a change in the structure of its DNA (genotype), but a change in the repertory of genes being expressed (epigenesis). Thus, the somatic mutation and epigenetic theories provide alternative ways of explaining the stability of the phenotype of neoplastic cells.

Pathologists have asked the question, 'What makes neoplastic cells different from their normal counterparts?' They concluded that all the properties of neoplasms were found in at least some types of normal cells: proliferation, invasiveness, and the ability to metastasize are expressed during early normal development, and even during adulthood, by various cell types. Thus, is there any entity that may be called *the* neoplastic cell? Or is neoplasia, instead, an emergent phenomenon resulting from a flaw among cells and tissues?

The administration of chemical carcinogens to animals that spontaneously developed certain neoplasias, increased tumor incidence and decreased tumor latency periods. This allowed the serial study of changes in tissue organization that preceded the development of a palpable or visible tumor. Interestingly, two very dissimilar experimental approaches emerged, one that was strictly reductionistic which adopted the premise that the deviation from normalcy was a cellular phenomenon caused by mutations, and another, integrative approach, which recorded in detail the subtle changes in tissue organization that preceded the establishment of the neoplasia. Those following the integrative approach, a minority, thought that since tissue organization was altered, carcinogenesis was due to the disruption of communications among cells and tissues. In short, carcinogenesis was interpreted as a cellular and mutational phenomenon by some, and as an integrative phenomenon involving disruptive interactions among cells and tissues by others.

## Germ-line mutations and carcinogenesis

Certain types of human cancer appear frequently among families; this suggested that genetic defects in the germ line predispose for the occurrence of cancer. Inbred strains of animals also revealed an inherited propensity to develop specific neoplasias. However, these germ-line mutations neither support nor rule out the somatic mutation theory, since there is no evidence that the mutation **directly** affects the somatic cells that will become neoplastic. Even among those holding conflicting views on carcinogenesis, there is no major disagreement regarding whether or not familial neoplasms are transmitted from parents to offspring through the germ line. The controversy is centered, instead, on whether there is a **direct** or just **indirect** correlation between germ-line mutations and the tumor phenotype. The latter option suggests that the mutated gene affects organizational phenomena at the tissular level.

In this context, it is worth recalling that the identification of the genetic defects underlying certain diseases has not always clarified how these genes result in the characteristic phenotype of the individual affected. For example, let us consider the Lesch-Nyhan syndrome. The main characteristics of this lethal syndrome are automutilation (children bite their fingers, hit their heads against hard objects, etc.), neurological signs (choreoathetosis, spasticity), and mental retardation. This disease is the result of a germ-line mutation in the enzyme hypoxanthine-phosphoribosyl transferase. The gene coding for this enzyme is located in the X chromosome. The disease appears in boys. Girls who carry a mutated gene and a normal allele of the gene do not develop the disease (they are just 'carriers'). These boys are asymptomatic at birth; the first signs of neurologic impairment appear at 3–4 months of age. So far, there is no explanation about how a defect on an enzyme that metabolizes purines results in these severe behavioral and neurological anomalies. Another interesting example in this regard is that of Down syndrome, due to a trisomy of chromosome 21. Patients suffer from mental retardation, malformations of the heart, and a characteristic 'mongoloid' phenotype. It is unknown how an extra copy of a normal chromosome results in the specific phenotype of this syndrome. The euphemism 'chromosomal imbalance' just covers over our current ignorance of this correlation. The identification of the genomic 'cause' does not provide 'mechanistic' clues of how the genotypic defect is translated into a phenotype when emergent phenomena are involved. In conclusion, it is unclear how germ-line mutations are responsible for carcinogenesis in humans and experimental strains of laboratory animals.

## Somatic mutations, cell proliferation, cell death, and the 'differentiation' hypotheses

The lacks of fit between data and the predictions of the mainstream somatic mutation theory of carcinogenesis were highlighted by several researchers in the second half of the 20th century. On the one hand, the somatic mutation theory does not predict how mutations would result in the expression of the neoplastic phenotype. On the other hand, the proponents of this theory assumed that the primary defect was the loss of proliferation control. However, the rate of proliferation of cells in neoplasms is not higher than that of cells in normal tissues. Moreover, the proliferation rate of cells in neoplasms of hormone-target organs is susceptible to the hormonal milieu in the host. For example, breast and prostate neoplasms regress when their trophic hormones, estrogens and androgens, respectively, are removed. This indicates that neoplasms are not necessarily autonomous entities. These behaviors cannot be readily explained by irreversible mutations.

A new twist was introduced within the somatic mutation theory to explain the fact that cells in neoplasias do not proliferate faster than their normal counterparts. About 25 years ago, apoptosis was first described to be as important a component as the generation of new cells to the regulation of cell numbers in tissues [3]. Regulation of apoptosis was also assumed to play a role in the rate of tumor growth. This notion was further extended into carcinogenesis by proposing that disabled genes involved in apoptosis may cause cancer [4].

In transplantation experiments, Pierce *et al.* observed that teratocarcinoma cells gave rise to 'differentiated' and non-tumorigenic tissue. They then postulated that the main defect resided in an altered control of differentiation, and showed that the assumption of the somatic mutation theory, 'once a cancer cell, always a cancer cell' was not supported by data. Instead, they considered that epigenetic mechanisms were at work without elaborating at which level of organization the defect was taking place [5]. Farber also pointedly criticized the assumption that the primary defect in carcinogenesis was on the control of cell proliferation [6].

Harris, a pioneer of somatic cell genetics, considered carcinogenesis to be a cellular phenomenon whereby the genome of the affected cell must be disrupted by lack-of-function mutations. In his own words: 'As things now stand, it appears that the key cellular events determining malignancy are heritable losses of function and, in particular, loss of the ability to complete specific patterns of differentiation. This may well be true not only for genetic lesions involving tumour suppressor genes, where the evidence is in some cases compelling, but also for mutated oncogenes. The two great peaks that somatic cell geneticists have long been attempting to scale, cancer and differentiation, seem to have merged into one.' [7].

It has also been suggested that interactions between parenchymal cells and their surrounding extracellular matrix play a critical role in neoplasia. Bissell and associates proposed that normal tissue architecture, which is affected by the extracellular matrix, is a high order repressor of the malignant phenotype of mutated cells [8]. In this instance, Bissell adds a new layer of complexity to the traditional notion that somatic mutations are the cause of neoplasia.

In summary, the hypotheses positing that carcinogenesis is an error in either cell proliferation, cell death, or cell differentiation, or a combination of these three processes, still adopt the notion that mutations are causal agents, and that neoplasia is a cellular phenomenon.

## Are carcinogenesis and neoplasia cellular or tissue-based phenomena?

In the following chapters, we will explore the subject of carcinogenesis. We will leave aside a discussion of the behavior of established neoplasias, and of areas that are intimately linked to them. For example, we will not address the need of neoplastic tissues to recruit an appropriate blood supply (neovascularization or angiogenesis). The process of metastasis, while also important to the biology of cancer, will not be discussed either.

We will examine the premises and tenets posited by the somatic mutation theory, and its subordinate hypotheses, and review the data gathered to explore them. We will also propose an alternative theory we call the *tissue organization field theory of carcinogenesis and neoplasia* which posits that neoplasias are emergent phenomena resulting from a flawed interaction among cells and tissues.

In a cogent and impassioned analysis in 1962, Smithers raised most of the pertinent questions regarding the somatic mutation theory, or cytologism, as he called it. He offered a series of suggestions to remedy the impasse in this highly controversial field [9]. Among several remarkable passages, we selected one that is both intriguing and eminently valid as we approach the beginning of the next century; 'Observation has produced too many incompatibilities, and a vast research effort too little support, for conventional cancer theory to hope to hold its place much longer'. The verdict of time has not been kind to Smithers, or others, who ventured into the realm of predictions in the field of cancer. With an additional four decades of intense efforts designed to vindicate the somatic mutation theory, the incompatibilities have been compounded and the uncertainties have multiplied. Has the time for a paradigmatic change in understanding carcinogenesis finally arrived?

## References


1. **Robbins, S.L.** (1967) *Pathology,* 3rd Edn. W.B. Saunders Company, Philadelphia, PA, p. 105.
2. **Foulds, L.** (1969) *Neoplastic Development.* Academic Press, New York.
3. **Wyllie, A.H., Kerr, J.F. and Currie, A.R.** (1980) Cell death: the significance of apoptosis. *Int. Rev. Cytol.* **68**: 251–306.
4. **McKenna, S.L. and Cotter, T.G.** (1997) Functional aspects of apoptosis in hematopoiesis and consequences of failure. *Adv. Cancer Res.* **71**: 121–164.
5. **Pierce, G.B., Shikes, R. and Fink, L.M.** (1978) *Cancer: A Problem of Developmental Biology.* Prentice-Hall, Englewoods Cliffs, NJ.
6. **Farber, E.** (1995) Cell proliferation as a major risk factor for cancer: a concept of doubtful validity. *Cancer Res.* **55**: 3759–3762.
7. **Harris, H.** (1995) *The Cells of the Body: A History of Somatic Cell Genetics.* Cold Spring Harbor Laboratory Press, Plainview, NY, p. 234.
8. **Weaver, V.M., Peterson, O.W., Wang, F., Larabell, C.A., Briand, P., Damsky, C. and Bissell, M.J.** (1997) Reversion of the malignant phenotype of human breast cells in three-dimensional culture and in vivo integrin blocking antibody. *J. Cell Biol.* **137**: 231–245.
9. **Smithers, D.W.** (1962) Cancer: an attack of cytologism. *Lancet* 493–499.

## Chapter 8

# The Enormous Complexity of Cancer

"... no theory of cancer—or of biology—is acceptable unless it comprehends neoplasia as one of the possible consequences of biological organization."

*L. Foulds (1969) Neoplastic Development. Academic Press, London (Preface)*

"Some investigators are fond of saying 'What we need is more facts.' The truth is that we already have more 'facts' than anybody knows what to do with. Experimental analysis has produced an alarming mass of empirical facts without providing an adequate language for their communication or effective concepts for their synthesis."

*L. Foulds (1969) Neoplastic Development. Academic Press, London (Preface)*

## Introduction

To any dispassionate observer of the cancer scene, the proverbial light at the end of the tunnel for the comprehension of cancer has not become brighter as time has gone by. In other words, many technological and even conceptual milestones have neither significantly improved our understanding of carcinogenesis, nor have they vastly affected the long-term survival rate of cancer patients. This fact is acknowledged by those who champion the current prevalent approach and set policy on how to continue fighting the War on Cancer which was declared in the early 1970s. For instance, the current Director of the National Institutes of Health, and 1989 Nobel prize laureate for his discovery of oncogenes, Harold Varmus, along with Robert A. Weinberg, candidly recognized that '... Unfortunately, the more refined cancer diagnoses now possible for tests for mutant genes have not yet provided clinically useful information in most situations ... Even for tumors in which certain oncogenic mutations are frequently encountered, ... informative correlations have not yet emerged—which is especially disappointing given the precision and elegance of these diagnostic tests . . .' Even the widespread use of chemotherapy is recognized for its shortcomings by these researchers: '. . . Despite its sometimes spectacular successes with certain cancers, the overall verdict on chemotherapy is not favorable, and the five year survival rates for most tumors have not improved significantly in the years since drugs were first widely used . . .' [1].

Another recent, largely unchallenged assessment of this collective failure by recognized epidemiologists and long-term observers of the field started with the following sentence: 'Despite decades of basic and clinical research and trials of new therapies, cancer remains a major cause of morbidity and mortality' [2]. Bailar and Gornick went on to highlight their long-standing position, namely, that it is now time to concentrate efforts on preventive measures because of the largely unproductive therapeutic attempts, and the lack of a basic research strategy anchored on a solid theoretical background. Regardless of how each stakeholder interprets these remarks, all of them may, nonetheless, still simply ask 'Why?' **Yes**, why is it that the conceptual understanding of cancer has escaped the concerted efforts of outstanding, devoted experimentalists and theoretical scientists working in diverse, but complementary disciplines for more than a century?

In Part I of this book, we presented a theory on control of cell proliferation, and discussed experimental evidence favoring the notion that *proliferation* is a built-in property of all living cells. In Part II, we examine how this reappraisal affects current views on carcinogenesis and neoplasia. The proposed switching of paradigms that we developed in Part I aims at providing the 'adequate language' that Leslie Foulds referred to in his 1969 book regarding the place that cancer occupies in biology [3].

### *Definitions in the dark*

Two criteria will frame the discussion of this subject. The first is that *proliferation* is the default state of *all* living cells, and the second is that the subject of what is broadly called cancer should be examined from a hierarchical perspective.

One of the most frustrating aspects in dealing with cancer, for specialists and lay people alike, is the lack of a concise and rigorous vocabulary. The current vocabulary has its pitfalls and therefore the narrative becomes, at times, messy and difficult to follow. From the onset, it should be flatly stated that no single, comprehensive definition of cancer can satisfy all aspects of this disease. This is a tacit acceptance of a lack of knowledge on its pathogenesis. We will use the term *cancer* when referring to the disease that affects humans. There are two main topics that will be kept conceptually separated: they are *carcinogenesis* and *frank neoplasia*. Carcinogenesis refers to the initial steps of tumor formation: this is a dynamic process that may stop temporarily or permanently, revert to normality, or progress to frank neoplasia. Neoplasias may invade, metastasize, and/or, albeit rarely, regress. The term *tumor* will be used sparingly because it is less accurate than neoplasia; tumors are bulging masses of tissue that may or may not be linked to a neoplasia (e.g. an inflammatory process may also appear as a 'tumor').

## Hierarchical levels of complexity in cancer

In our view, four major hierarchical levels encompass the diverse domains inherent to cancer. These levels are, first, a *social level* which includes the patient's relationship with his/her family, doctors, and social contacts. This level also encompasses cancer epidemiology and the bureaucracy that deals with cancer research. Second, an *organismal level*, in which cancer represents the complex disease at the level of the individual (human or experimental animal). Third, an *organ/tissue hierarchical level*, where the pathologist defines the neoplastic state by describing the characteristics of the affected *tissue* under the magnifying power of the light microscope. Finally, a *cellular/subcellular hierarchical level*, where neoplastic cells are operationally defined as cells that have been derived from an actual neoplasia present in either a patient or an experimental animal. Following a cartesian strategy, over decades, researchers have disassembled neoplasias to investigate what was unique about their cells, and within these, what was unique about their organelles and molecules. This is a reflection of the success among scientists and medical practitioners of Virchow's cell pathology paradigm. In the process of 'reducing' the object of analysis much has been learned about the parts, while the understanding of the whole has not fared so well.

### *The social level of hierarchical complexity of cancer*

This level encompasses epidemiological studies that have revealed links among environmental exposure, lifestyles, heredity and cancer. This level also touches what has been vaguely called the politics of cancer. It includes the appropriation of funds to promote cancer research, the financial impact on educational and research-oriented institutions, and the effort of activists to influence the allocation of those funds. The term 'cancer industry' has been coined to describe the multiple interests that revolve around what has become 'the American disease' based on the frightening number of patients in our country who become ill and eventually die from cancer. This sociological level of complexity and its influence in cancer research is beyond the scope of this book [4, 5].

### *The organismal hierarchical level of cancer*

Prokaryotes (such as bacteria) and unicellular eukaryotes (such as yeast, or amoeba) do not develop neoplasia. Only multicellular organisms do. Species where neoplasia develops 'spontaneously' and represents a sizable problem are mostly restricted to humans and the utilitarian species and breeds that humans have domesticated. Left to the vagaries of natural selection in the wild, metazoa are rarely victims of neoplasias. Only recently, has an increased incidence of neoplasia been noted in species living in polluted habitats (fish, whales) [6, 7].

Neoplasias are the underlying physical entities responsible for the multiple symptomatology of the cancer disease. The formation of visible or palpable tumors usually takes a long period of time; its length is proportional to the characteristic lifespan of the species. In humans, this period may last several decades. Only those neoplasias that increase their cell number over time occupy the attention of the patient or the medical profession. The natural history of these carcinogenic processes has shown that they may either progress, not increase in size, or regress to normality [8]. This multiple, diverging fate encouraged commentators to claim that cancer is not a single disease but many diseases.

From a clinical perspective, cells in a neoplasia not only ignore the proliferation control that their normal counterparts are subjected to, but they may also invade adjacent tissues. Neoplastic cells may enter lymph and/or blood vessels and migrate to other organs, where they establish *metastases*. Invasiveness and metastases are the principal phenomena responsible for the death of the cancer patient. Incidentally, neoplastic cells migrate in ways similar to those used by normal cells in multicellular organisms. Migration of normal cells is especially obvious during early stages of development but also happens during adulthood; for example, blood cells move from their tissue of origin (the bone marrow) into the bloodstream and from the vascular compartment to most tissues of the organism where they play their diverse physiological roles. Hence, motility (invasion) through tissues or blood is not a unique, novel property of neoplastic cells, but the reacquisition of a biological capability that was exercised during earlier developmental stages. Cell motility, as proliferation, may also be considered as a built-in property of all cells [9].

In addition to the organismal dislocations generated by the increased size or invasiveness of the neoplasias, we humans are also affected psychologically by cancer when the issue of our mortality is brought forward. We refer readers to specialized books and reviews on this topic [10, 11].

### *The tissue level of complexity in cancer*

Clinicians, radiologists and surgeons suggest the diagnosis of cancer. The pathologist provides the final verdict of whether the tissue being examined is neoplastic. The tool he/she uses to make this decision is an uncomplicated light microscope. What the pathologist identifies is a pattern of tissue organization. The neoplastic tissue may grossly or subtly differ from a normal pattern. In the initial stages of carcinogenesis, this is an admittedly subjective exercise [12, 13]. Usually, pathologists describe a static view of a neoplasia once it has developed for years, and then combine their accumulated experience of how a histological pattern matches with the natural history of the disease. That is, the pattern read today is used to prognosticate, based on experience, what the most probable outcome will be in the future of a particular patient.

Experimentally-induced precursor lesions and frank neoplasias in laboratory animals are also defined according to histological criteria. Most importantly, no single cytoplasmic or nuclear feature in any cell is sufficient to unequivocally diagnose a neoplasia [14]. From this and other considerations to be reviewed below, we have aligned ourselves with others who proposed that carcinogenesis is a process inherent to the tissue hierarchical level of complexity [8, 15–19].

#### *A revived theory of carcinogenesis and neoplasia*

We posit that in normal, adult, multicellular organisms there are discrete *units of tissue maintenance and/or organization*; histologically, they comprise the parenchyma and the stroma of an organ. During embryogenesis, adjacent stroma and epithelia exert instructive influences on each other resulting in organ formation. This complex web of interactive signaling continues throughout the lifespan of the individual. We further postulate that these units of tissue maintenance and/or organization are tridimensional and carry positional and historical information. They maintain the normal architecture of all organs and guide tissue turnover, remodeling and healing through a dynamic process. These units of tissue maintenance and/or organization are present in all organs and are the ultimate targets of carcinogenic agents. This theory posits that teratogenesis, developmental tumors, and 'spontaneous' and induced carcinogenesis, occur because of miscommunication among cells and tissues.

In 'spontaneous' and agent-mediated carcinogenesis there is disruption of the normal interactions that take place among cells in the parenchyma and subjacent stroma of an organ. This disturbance results in functional and structural changes in the affected tissue/organ. Individual cells within these lesions recognize positional miscues among themselves and their neighbors. For instance, cells in the parenchyma may increase their proliferation rate (*hyperplasia*) and/or show the two elements of *dysplasia* which are a less orderly organization than that found in the normal tissue of origin, and an increased production of new cells. Another type of response to a perturbation of the maintenance of normal organization is illustrated by the appearance of *metaplasias*, whereby adult tissues express their ability to become other tissue types (e.g. from simple cuboidal to stratified squamous epithelia). These metaplasias occur frequently at the point where two segments of hollow organs join each other (say, the esophagus with the stomach, or two regions of the uterine cervix) [20].

Further deterioration of the tissue organization and/or maintenance field leads to what has conventionally been called carcinoma *in situ* (in organs that have an epithelial lining). If the disturbance ceases, the damaged tissue may gradually reverse course and regain its normal architecture and cellular phenotypes. If, instead, the injurious circumstances persist, progressive tissue deterioration may ensue and the condition may reach a frank neoplasia. Restraining cues emitted by other less damaged or normal cells laying within the field may not be effective. Still, homeostatic mechanisms operating at the organismic level during adulthood through signals such as hormones and inhibitors of cell proliferation still may restrain the built-in capacity of these cells to proliferate. In short, we conclude that carcinogenesis and neoplasias are the result of the disruption of the above-mentioned tissue organization fields. If the affected unit could be restored to some degree of 'normality', the progression to ever greater cellular autonomy could be delayed, or even totally reversed. In *Figure 8.1*, a general, schematic view on experimental carcinogenesis is presented. The theory acknowledges the premise that the default state of all cells is *proliferation* and that the tissue organization and/or maintenance field units are the ultimate target of carcinogens.

### The cellular/subcellular hierarchical level

For almost a century now, the view that carcinogenesis takes place at the cellular and subcellular levels has been a prevalent one. The *somatic mutation theory of carcinogenesis* can be formulated as follows: 'one or more mutations on the genome of *a somatic "normal" cell* will render it unable to effectively control its proliferation; as a consequence, such a cell will become *a neoplastic cell.*' The implicit premises of this hypothesis are: (1) carcinogenesis originates at the single cell level; (2) mutations must result in an increase of the proliferative rate of the *neoplastic cell*; and (3) the default state of metazoan cells is *quiescence*. Hence, mutations that may induce neoplastic behavior should be of the 'gain-of-function'

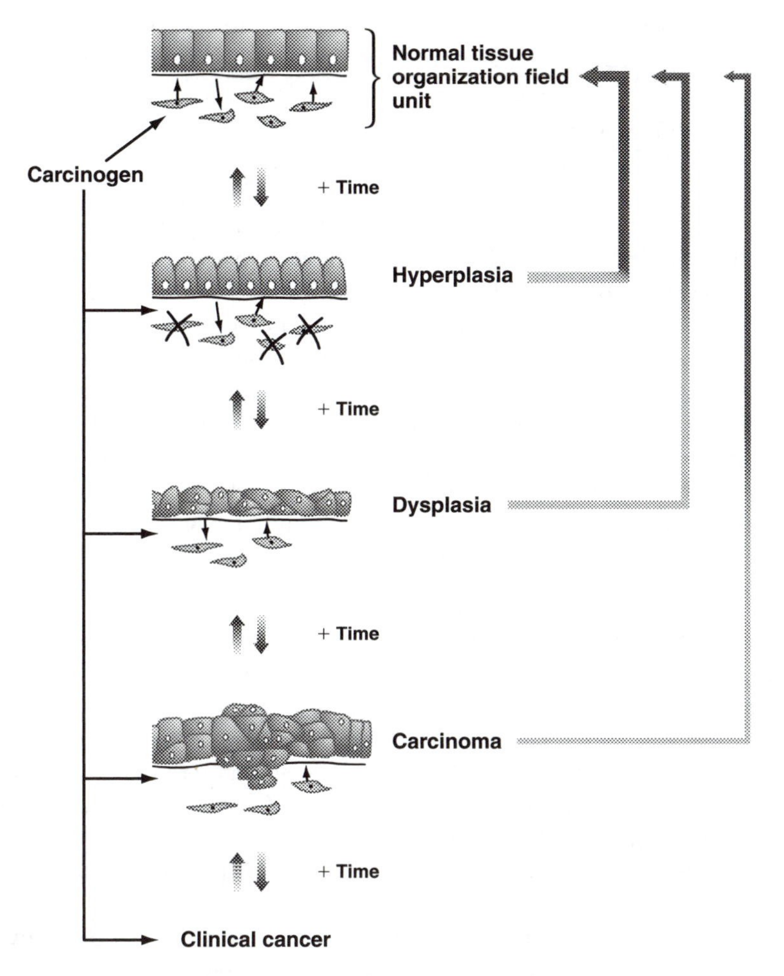


**Figure 8.1.** Schematic representation of the process of experimental carcinogenesis according to the *tissue organization field theory.* A carcinogen acts on the intact field units. The phenotypic alteration is mainly observed in the epithelial lining adjacent to the equally affected stroma. If the carcinogen-induced damage in the tissues is not profound, reversibility of the phenotypic epithelial defects (hyperplasia, dysplasia, metaplasia) is facilitated by removing the carcinogen. This general scheme can accommodate explanations for experimental and 'spontaneous' carcinogenesis.

type; their effect would be to promote the proliferative ability of cells. These mutated genes are called oncogenes. Conversely, 'loss-of-function' mutations would inactivate inhibitory signals coded by anti-oncogenes. This latter notion represents one of the many *ad hoc* additions to the original core of the somatic mutation theory. We will deal more extensively with these additions in Chapter 9.

It has been clearly stated that mutated oncogenes would code for growth factors, their receptors, proteins in the signal transduction pathway, components of the cell cycle machinery, and other assorted targets. The products of these mutated oncogenes would stimulate cells stuck in *quiescence* to proliferate and, thus, generate a tumor [21].

*Inferences from cell culture to the tissue level of complexity. 'Transformation'*
The idea that carcinogenesis originates in *a single mutated cell* represents a significant technical disadvantage to the experimentalist. It is practically impossible to work on a single cell in an essentially dynamic process like carcinogenesis. In order to overcome this limitation, cell populations have been grown *in culture* under the tacit assumption that they are homogenous units, and that data collected using this model can be 'reduced' down to the individual cell level. Beginning in the late 1950s and well into the 1960s, experiments designed under this rationale have been used to test the effect of chemical carcinogens and radiation. Following exposure, the putatively mutated cells would change their proliferative behavior and their phenotypic appearance *in culture*; this has been defined as 'transformation'.

These operationally defined 'transformed' cells were considered as equivalent to truly neoplastic cells present in human tumors and in experimentally induced tumors in animals. However, when these cells were tested for their ability to form tumors in susceptible hosts, it became plain that this correlation contained many exceptions. To begin with, since the 1940s, it was reported that 'transformation' occurred 'spontaneously' in the absence of any putative carcinogen in the culture medium [22–24]. Notwithstanding, beginning in the 1960s and 70s, chemical carcinogens [25] and ionizing radiation [26, 27] were reported to 'transform' so-called normal cells into neoplastic cells. These data were interpreted as favoring the notion for a **direct** mutagenic effect of carcinogens on individual cells.

Meanwhile, in the late 1970s and early 1980s, a number of researchers suggested that the transferring of DNA fragments from cancer cells into 'normal' mouse cells resulted in a neoplastic phenotype [28]. These and additional experiments in which oncogenes were used, were interpreted as evidence that the DNA from cancer cells and oncogenes carried the information necessary to confer a neoplastic behavior (gain-of-function) [29]. This interpretation was criticized, first, because the 'transformed' phenotype remained fixed even when the oncogene was lost by these cells; second, the 'normal' cells used in these experiments usually undergo 'spontaneous' transformation; and third, neoplastic transformation involved the loss of genetic components as revealed by chromosome aberrations different from those present in their parental cells [30, 31].

Starting in 1980, several researchers began to question the proposed direct effect of physical and chemical carcinogens on 'transformation' *in culture* as being due to mutational events [32, 33]. Moreover, new data suggested that the phenotype and even the ability to generate tumors by these 'transformed' cells could be reverted to normalcy by just modifying the culture conditions under which those cells grew [34, 35]. These data suggested, instead, that transformation was a hereditary, epigenetic, highly variable phenomenon. Cells showing a transformed phenotype could be switched back to a normal phenotype, and vice versa, just by modifying nutritional conditions in the experimental setting of a plastic dish. The capacity of these cells to develop as tumors when injected into susceptible hosts was also dependent on the conditions under which these cells were kept *in culture*. Rubin and associates coined the term 'progressive state selection' to characterize this *in-culture* based phenomenon [36].

Remarkably, none of the transformation experiments *in culture* recapitulated the steps that tissues follow when engaged in the process of carcinogenesis. That is, hyperplasia, dysplasia, metaplasia, and carcinoma *in situ* are phenomena occurring at the tissue level of hierarchical complexity. They only happen in epithelial tissues in animals during the process of spontaneous or experimentally induced carcinogenesis. Dissociating tissues and placing their now disconnected cells *in culture* represents an extreme disruption of the organization fields from which they originate. This 'demergent' state may lead by itself to the generation of neoplasia by merely freeing cells from their *in situ* positional restraints. This *in-culture* approach does not reproduce organismal phenomena. It represents, instead, an artifact equivalent to reversing the evolutionary process (from metazoa to unicellularity).

### *Conflicts within the somatic mutation theory of carcinogenesis*

The use of cells *in culture* transplanted into susceptible hosts pertains to the behavior of frank neoplastic cells, and as mentioned above, it does not recapitulate the process of carcinogenesis. In the 1960s, somatic cell geneticists explored the notion of whether 'malignancy' was a dominant or recessive trait [30]. Mice and human cell lines derived from neoplasias, and even from 'normal' cells, kept *in culture* for many years, were able to develop into tumors when injected into susceptible hosts. These 'malignant' cells were hybridized (fused) *in culture* with cells that did not produce tumors. When injected subcutaneously, these somatic cell hybrids failed to develop tumors in mice shortly after the hybridization took place. However, if the somatic hybrid cells were kept in culture conditions through additional passages, they began to randomly shed chromosomes. Next, it became apparent that the loss of certain chromosomes from the hybrid cells correlated with the recovery of their ability to develop as tumors in animals. In contrast to the notion that 'malignancy' was a gain-of-function or dominant character [1, 28, 29], the interpretation of these experiments suggested that 'malignancy' was a loss-of-function or recessive trait. These conflicting conclusions within the same theory suggest a lack of relevance of these data to explain the process of carcinogenesis.

### *The Drosophila file*

Development of tumors in *Drosophila* unambiguously link a germ-line mutation to the emergence of a neoplasia. Having characterized the mutated gene, however, has shed little light on why mutations in this gene result in the formation of a tumor. Neuroblastomas appear in larvae carrying homozygous mutations of a gene called *lethal giant larva-2* (*lgl-2*) [37]. The normal gene codes for an intracellular, cytoskeleton-associated protein expressed in the early embryo, long before the morphogenesis of the nervous system takes place. Injection of the *wild*-type allele into homozygous mutated embryos, but not later, results in normal flies [38].

From the *somatic mutation theory* perspective, the difficulty remains in trying to understand how the affected gene resulted in a neoplasia since this protein does not appear to have a direct role on the control of the proliferation of neuroblasts. The lack of an explanation provided by data gathered at hierarchical levels lower than those in which carcinogenesis takes place, may be better understood by the following metaphorical example: '[I]f you want to know why traffic jams tend to happen at a certain hour every day, you will still be baffled after you have painstakingly reconstructed the steering, braking and accelerating processes of the thousands of drivers whose various trajectories have summed to create those traffic jams.' This pursuit is what our colleague at Tufts, Daniel C. Dennett, an unabashed reductionist himself, called 'preposterous reductionism' [39].

When adopting the premises of the tissue organization field theory of carcinogenesis, one may postulate that the timely expression of *lgl-2* gene product is crucial to the communication among the cells responsible for the normal neural development in *Drosophila* at the larva stage. A loss-of-function mutation would affect the correct reading of cues to finalize the development of neural tissues during organogenesis. The fatal tumor would be the outcome of this misalignment.

## Conclusions

We have argued in this chapter that carcinogenesis is a defect inherent to the organ/tissue hierarchical level of complexity. The tissue organization field theory of carcinogenesis posits that units of organization and/or maintenance are operational in tissues and organs throughout the lifespan of multicellular organisms. These units are the targets of experimental carcinogens.

By applying a hierarchical approach to understanding the complexity of cancer, and by adopting the notion that carcinogenesis is a process that affects morphogenetic fields in early development, and their equivalents in adult multicellular organisms, we believe that a more rigorous understanding of cancer—*and* of biology—may now be closer at hand. We are confident that current available techniques and novel approaches used in developmental genetics under the premises and theory outlined above offer a more realistic opportunity to understand cancer and to manage it better. Further analysis of the merits of competing theories of carcinogenesis will be presented in the next chapter.

## References

1. **Varmus, H.E. and Weinberg, R.A.** (1992) *Genes and the Biology of Cancer.* Scientific American Library, New York, p. 123.

2. **Bailar, J.C. III and Gornick, H.L.** (1997) Cancer undefeated. *N. Engl. J. Med.* **336**: 1569–1574.

3. **Foulds, L.** (1969) *Neoplastic Development.* Academic Press, New York.

4. **Fujimura, J.** (1996) *Crafting Science.* Harvard University Press, Cambridge, UK.

5. **Proctor, R.N.** (1995) *Cancer Wars.* Basic Books, New York.

6. **Colborn, T., Dumanoski, D. and Myers, J.P.** (1995) *Our Stolen Future.* Penguin Books, New York.

7. **Steingraber, S.** (1997) *Living Downstream.* Addison-Wesley, Reading, pp. 118–147.

8. **Clark, W.H.** (1995) The role of tumor progression in prevention of cancer and reduction of cancer mortality. In: *Cancer Prevention & Control* (eds P. Greenwald, B.S. Kramer and D.L. Weed). Marcel Decker, New York, pp. 135–159.

9. **Buss, L.W.** (1987) *The Evolution of Individuality.* Princeton University Press, Princeton, NJ, pp. 69–117.

10. **Sontag, S.** (1990) *Illness as Metaphor and AIDS and its Metaphors.* Doubleday, New York.

11. **Husebo, S.** (1993) Why are people afraid of cancer? In: *New Frontiers in Cancer Causation* (ed. O.H. Iversen). Taylor & Francis, London, pp. 409–416.

12. **Rosai, J.** (1991) Borderline epithelial lesions of the breast. *Am. J. Surg. Pathol.* **15**: 209–221.

13. **Henson, D.E. and Albores-Saavedra, J.** (1993) Introduction. In: *Pathology of Incipient Neoplasia* (eds D.E. Henson and J. Albores-Saavedra). Prentice-Hall, Englewoods Cliffs, NJ, pp. 1–7.

14. **Koss, L.G.** (1992) *Koss's Diagnostic Cytology and its Histopathologic Bases,* 4th Edn. J.B.Lippincott Co., Philadelphia, PA, pp. 126–147.

15. **Waddington, C.H.** (1935) Cancer and the theory of organizers. *Nature* **135**: 606–608.

16. **Tarin, D.** (1972) Tissue interactions and the maintenance of histological structure in adults. In: *Tissue Interactions in Carcinogenesis* (ed. D. Tarin). Academic Press, London, pp. 81–94.

17. **Orr, J.W.** (1958) The mechanism of chemical carcinogensis. *Br. Med. Bull.* **14**: 99–101.

18. **Smithers, D.W.** (1962) Cancer: an attack of cytologism. *Lancet* 493–499.

19. **Rubin, H.** (1985) Cancer as a dynamic developmental disorder. *Cancer Res.* **45**: 2935–2942.

20. **Rao, M.S. and Reddy, J.K.** (1996) Cell and tissue adaptations to injury. In: *Cellular and Molecular Pathogenesis* (ed. A.E. Sirica). Lippincott-Raven, Philadelphia, PA, pp. 57–78.

21. **Alberts, B., Bray, D., Lewis, J.G., Raff, M., Roberts, K. and Watson, J.D.** (1994) *Molecular Biology of the Cell,* 3rd Edn. Garland Publishing, New York, pp. 1280–1281.

22. **Earle, W.R.** (1943) Production of malignancy in vitro. IV. The mouse fibroblast cultures and changes seen in the living cell. *J. Natl Cancer Inst.* **4**: 165–212.

23. **Gey, G.O.** (1955) Some aspects of the constitution and behavior of normal and malignant cells maintained in continuous culture. *Harvey Lect.* **50**: 154–229.

24. **Sanford, K.K.** (1965) An attempt to induce malignant transformation of cells in vitro by intermittent anaerobiosis. *J. Natl Cancer Inst.* **35**: 719–725.

25. **Berwald, Y. and Sachs, L.** (1965) In vitro transformation of normal cells to tumor cells by carcinogenic hydrocarbons. *J. Natl Cancer Inst.* **35**: 641–661.

26. **Terzaghi, M. and Little, J.B.** (1976) X-radiation induced transformation in a C3H mouse embryo-derived cell line. *Cancer Res.* **36**: 1367–1374.

27. **Borek, C. and Sachs, L.** (1968) The number of cell generations required to fix the transformed state in X-ray-induced transformation. *Proc. Natl Acad. Sci. USA* **59**: 83–85.

28. **Shih, C., Shilo, B.Z., Goldfarb, M.P., Dannenberg, A. and Weinberg, R.A.** (1979) Passage of phenotypes of chemically transformed cells via transfection of DNA and chromatin. *Proc. Natl Acad. Sci. USA* **76**: 5714–5718.

29. **Weinberg, R.A.** (1983) A molecular basis of cancer. *Sci. American* **249**: 126–142.

30. **Harris, H.** (1985) Suppression of malignancy in hybrid cells: the mechanism. *J. Cell Sci.* **79**: 83–94.

31. **Harris, H.** (1995) *The Cells of the Body: A History of Somatic Cell Genetics.* Cold Spring Harbor Laboratory Press, Plainview, NY, pp. 211–247.

32. **Kennedy, A.R., Fox, M., Murphy, G. and Little, J.B.** (1980) Relationship between X-ray exposure and malignant transformation in C3H10T1/2 cells. *Proc. Natl Acad. Sci. USA* **77**: 7262–7266.

33. **Little, J.B.** (1994) Changing views of cellular radiosensitivity. *Radiation Res.* **140**: 299–311.

34. **Paquette, B., Wagner, R. and Little, J.B.** (1996) In vitro reversion of transformed phenotype in mouse C3H-10T1/2 cells: modification in genomic 5-methylcytosine content. *Int. J. Oncol.* **8**: 727–734.

35. **Rubin, H.** (1992) Cancer development: the rise of epigenetics. *Eur. J. Cancer* **28**: 1–2.

36. **Rubin, A.L., Sneade-Koenig, A. and Rubin, H.** (1992) High rate of diversification and reversal among subclones of neoplastically transformed NIH 3T3 clones. *Proc. Natl Acad. Sci. USA* **89**: 4183–4186.

37. **Gateff, E. and Schneiderman, H.A.** (1969) Neoplasms in mutant and cultured wild-type of Drosophila. *Natl Cancer Inst. Monographs* **31**: 365–397.

38. **Mechler, B.M., Strand, D., Kalmes, A., Merz, R., Schmidt, M. and Torok, I.** (1991) Drosophila as a model system for molecular analysis of tumorogenesis. *Environ. Health Perspect.* **93**: 63–71.

39. **Dennett, D.C.** (1995) *Darwin's Dangerous Idea.* Simon & Schuster, New York.

## Chapter 9

# Facts and Fantasies in Carcinogenesis

"Cancer cells originate from normal body cells in *two* phases. The first phase is the irreversible injuring of respiration. Just as there are many remote causes of plague—heat, insects, rats—but only one common cause, the plague bacillus, there are a great many remote causes of cancer—tar, rays, arsenic, pressure, urethane—but there is only one common cause into which all other causes of cancer merge, the irreversible injuring of respiration.

The irreversible injuring of respiration is followed, as the second phase of cancer formation, by a long struggle for existence by the injured cells to maintain their structure, in which a part of the cells perish for lack of energy, while another part succeed in replacing the irretrievable lost respiration energy by fermentation energy. Because of the morphological inferiority of fermentation energy, the highly differentiated body cells are converted by this into undifferentiated cells that grow wildly—the cancer cells . . ."

*Otto Warburg (1956) On the origin of cancer cells. Science* **123***: 309–314*

"How cancer develops is no longer a mystery. During the past two decades, investigators have made astonishing progress in identifying the deepest bases of the process—those at the molecular level. These discoveries are robust: they will survive the scrutiny of future generations of researchers, and they will form the foundation for revolutionary approaches to treatment . . ."

*Robert A. Weinberg (September 1996) Scientific American, pp. 62–70*

## Introduction

Researchers know *how* to generate tumors in animals when applying well-defined experimental protocols. What we do not know is *why* this happens and how it relates to the way cancer develops in humans. Researchers have proposed theories to explain these unknowns. Two views recur throughout this century to explain carcinogenesis: the *somatic mutation theory* and its variants, and the *tissue organization field theory*.

### *The somatic mutation theory of carcinogenesis: where is it coming from?*

What has been the rationale to promote the somatic mutation theory of carcinogenesis? To answer this question will require us to go back two centuries. In the 18th and 19th centuries, biologists operated in the midst of two irreconcilable world views. The vitalist view proposed a 'vital force' to explain biological phenomena, whereas the physicalist–mechanicists insisted that everything could be explained by the laws of physics and chemistry, and viewed the organism as a machine. When the vitalists studied early embryogenesis, they realized that the development of organs and limbs could not be explained in physicalist terms by invoking changes in potential and kinetic energies or by the movement of particles. The physicalist–mechanicists then proposed that all parts of the adult existed in the early embryo (preformationists), while the vitalists believed that the adult parts appeared as a product of development, and moved on to describe how the new parts were formed (epigenesis). Ernst Mayr made the point that if one replaced the terms 'genetic program' for 'vital force', the vitalists' writings would make perfect sense today [1].

The concept of mutation as the cause of neoplasia is attributed to Boveri [2] who, in 1914, explained his view about how *a* neoplastic cell may originate from *a* normal cell. Since then, the implication has been that neoplasias are of monoclonal origin, that is, all the cells of the neoplasia originated from a single cell. Once Mendel's laws of inheritance were re-discovered around 1900, Sutton and Boveri proposed the chromosome theory that posited that chromosomes were the material substrate of inheritance. Genes were thus orderly located in the chromosomes. Boveri took the view that a neoplastic cell resulted from 'abnormal mitosis'. Chromosomes would be unevenly distributed in their daughter cells, and the cells that survived would transmit their anomaly to their own daughters. He specifically stated ' . . . The cell of a malignant tumor is accordingly . . . a cell with a definite abnormal chromatin-complex' [2].

In the early 1940s, Avery and his associates demonstrated that DNA was the genetic material [3]. Watson and Crick, in 1953, solved the DNA structure which provided a molecular explanation for the precise transmission of genetic information from mother to daughter cells [4]. Before then, the consensus among biochemists, geneticists, and embryologists was that genes were made up of proteins. The complementary double-helix structure of the DNA molecule provided the opportunity to explore the mechanism of action of mutagens. Throughout this century, the somatic mutation theory of carcinogenesis remained conceptually intact despite the significant change in the final molecular target of the elusive mutation. In this context, the crucial question then turned out to be, how do carcinogens disrupt the 'inheritance factors' of Boveri, or the genes of the geneticists, to produce neoplasias?

Before the 1950s, it was impossible to grasp how the developmental program was genetically coded. This was accomplished with the marriage of molecular biology and information theory ('genetic program'). While the physicalist–mechanicists have remained loyal to their centuries-old tradition, by the 1930s, the vitalists' need to invoke a 'vital force' was superseded by a new paradigm, organicism. This explanatory system claims that at lower hierarchical levels (molecular), biological phenomena could be explained by the laws of physics and chemistry; however, at higher levels of organization (tissue, and above), the emergent and integrative phenomena could not be explained directly in the terms of physics and chemistry. For example, it is pointless to explain sexual reproduction or mate selection in molecular terms despite recognizing that matter is made up of molecules and atoms. Following this rationale, the concept that carcinogenesis is an emergent phenomenon resulting from abnormal tissue organization could not be formulated until embryologists created the concepts of organizers and morphogens to explain embryonal development; these concepts were established in the 1920s and 30s, mainly through the efforts of Spemann. Ironically, it was Boveri who first introduced the notion of morphogenetic fields.

The embryologists' efforts to explain development were thwarted, however, by a combination of technical and ideological factors. Eventually, the scientific credo of genetics, a paragon of linear thought (one gene–one enzyme), triumphed over that of embryology, which dealt with the complexity of interactions among cells and tissues [5]. Once the hegemony of genetics over epigenetics was established, and DNA was identified as the genetic material, genes acquired both a physical reality and a mystique, and so did mutations. The next step was to 'reduce' modern biology into the molecular biology of the cell.

Neoplasias are seen as the consequence of mutations in *a normal cell*, which would evolve into *a cancer cell*. It is intriguing to consider that while in the 1960s it became clear that somatic cells expressed different phenotypes while maintaining their genomes intact [6], three decades later a majority of researchers still consider a mutation(s) a *sine qua non* cause of cancer. Circumstantial evidence for a role of mutations in carcinogenesis was provided by the fact that many chemical carcinogens are mutagens; however, a significant number of them are not. This posed two questions: the first was, if some carcinogens are mutagens and some are not [7–9], how can one adduce that mutations are necessary? The second was, what gene(s) has/have to be mutated to produce a histological lesion with the physical and behavioral characteristics of a neoplasia? Answers to these questions are still missing.

### *The somatic mutation theory of carcinogenesis: to where has it led us?*

In the last three decades, researchers committed to the somatic mutation theory have thoroughly searched for mutations that could alter the ability of cells to proliferate. Their views on carcinogenesis and control of cell proliferation overlapped. As explained in Part I, the premise adopted by those working on the control of cell proliferation in metazoa has been that the default state was *quiescence*. Following this rationale, researchers explored the role of growth factors and their receptors. The assumption underlying these quests is that *a* quiescent cell that would normally enter the cell cycle when stimulated by growth factors, would acquire mutations in genes coding for growth factor receptors, and the signal transduction pathways that relay the growth factors signals into the cell. These activated gene products would then allow the cell to proliferate extemporaneously in the absence of growth factors. The proponents of the somatic mutation theory also appear to assume that the control of the cell cycle and that of cell proliferation are one and the same. This contradicts the notion of the existence of a default stage and of a switch that moves cells in and out of the cycle (from quiescence to proliferation). Thus, they posit that gain of function mutations on cell cycle effectors (oncogenes), or loss of function mutations on cell cycle inhibitory components (antioncogenes) may result in increased cell proliferation, and carcinogenesis.

Let us briefly examine the notion that cell cycle components may regulate cell proliferation. The cell cycle functions like a self-contained program. Its operation was compared with that of an automatic clothes washing machine [10]. Once the switch is turned on, the program is performed to completion. Sensors inside the machine read the water level, temperature, and so on, and determine when (not whether) the next step will be initiated. Thus, the 'control' is the decision made by turning on the switch. Once the switch is on, the program is performed inexorably. Deletion of the genes that carry on the information to perform the different steps of the cycle (cyclins, kinases, etc.) is lethal [11]. This set of genes was extensively studied in yeasts, where they were identified by the generation of temperature-sensitive mutants. At a permissive temperature the mutated gene is functional, and at the restrictive temperature the function of the mutated gene is switched off. Of course, at the restrictive temperature the affected cells are unable to proliferate.

In *Drosophila,* the knocking-out of the E2F1 transcription factor revealed that it is required for entry into the S phase. Experimental over-expression of this gene showed that it is not able to override the controls that determine whether and when a specific cell will proliferate [12]. In other words, it does not control cell proliferation. The proponents of the somatic mutation theory postulated that loss of function mutations of putative negative regulators of the passage from the $G_1$ to the S phase of the cycle (Rb (see below), p107, p130, etc.) may result in neoplasia because cells would undergo the cycle in the absence of stimulatory signals. However, in knockout mice, it was shown that at least early development occurs normally [11]. Cell culture experiments were carried out to demonstrate subtle alterations of the control of cell proliferation in cells from these knockout mice. However, it was found that in the Rb1 knockout ($Rb^{-/-}$) the $G_1$ phase was shortened while the S phase was prolonged, and thus the duration of the cell cycle remained similar to that of the *wild*-type cells [13]. Fibroblasts from p27 knockout mice (another putative negative regulator of the kinases involved in the passage from $G_1$ to S), responded to the inhibitory effect of TGF-β. This indicates that deletions of p27 are not able to override proliferation-inhibitory signals [14].

Support for the involvement of cell cycle components as signals or regulators of the entry of a cell into the cycle is based on a controversial reading of the experimental data [13, 15, 16]. From the theoretical perspective postulating the existence of a switch, this is not surprising. The on/off switch for entry into the cell cycle, or into quiescence, signifies that the signals are read and acted upon at the $G_0/G_1$ interphase, and that the cell cycle will proceed automatically once the switch is turned on (see Chapter 1). Control of cell proliferation is a process taking place at the cellular hierarchical level, chronologically removed from cell cycle control (subcellular hierarchical level). Again data coming from a lower level of organization seldom explain what is going on at the higher level of inquiry.

An enormous amount of data at the biochemical and molecular levels has been accumulated while exploring the somatic mutation theory. Nonetheless, like others before and since have discovered, Farber concluded that ' . . . despite the intensity of human and material resources that have been focused on cancer *per se,* the essential biochemical and genetic basis for the different major properties of cancers, such as autonomy of growth, invasion and metastasis, continue to elude the cancer researchers . . . ' [17]. However, other researchers still insist on portraying a dubious biological reality where most mysteries in cancer have been resolved, and just a few details are missing. These contradictory views breed confusion, anger and cynicism among patients, their relatives, their physicians, and the public at large. Who should one believe? Admittedly, a lot has been learned about cell cycle progression (cyclins, kinases, etc.), virus–host interactions, signal transduction, *et cetera.* This increased knowledge has illuminated important areas of cell biology, biochemistry, and genetics. Regrettably, however, those data are not germane to the understanding of carcinogenesis, proliferative autonomy, invasion, and metastasis. These are integrative phenomena involving interactions among cells, tissues, and organs, the understanding of which is unlikely to be enhanced by exploring intracellular biochemical and molecular pathways. The hoped-for results promised by looking inside the cell have not been delivered despite the long-term, heavy investment by all segments of the scientific community. However painful it may be, it should be acknowledged that the exploration of the somatic mutation theory may have run its usefulness as a guide to resolve the process of carcinogenesis.

### *The tissue organization field theory of carcinogenesis. Starting anew*

*The tissue organization field theory of carcinogenesis and neoplasia* states that carcinogens disrupt the flow of information between the stroma and the parenchyma and/or among cells within those tissues. The temporary or permanent effects of carcinogens on the intracellular structures and components, while variably deleterious to each of them, are not **directly** responsible for the development of a neoplasia. In the experience of Orr and his associates [18, 19], carcinogens act preferentially on cells in the stroma which however do not become neoplastic (see below). According to this theory, these carcinogen-altered cells in the stroma affect the progenitors of would-be neoplastic cells in the parenchyma. The altered epithelial cells are perceived as different by their neighbors, which in turn respond to the change in their immediate environment. This chain of miscommunication is a slow and subtle positive feedback of change that generates more change. Eventually, signals that restrain cells to express only the appropriate phenotype due to their position are weakened. This increases the probability of cells in this tridimensional field to express their built-in ability to proliferate. Hence, carcinogenesis and neoplasia may occur entirely through emergent (supracellular) phenomena once the signals that maintain normal organization are disrupted. Kaufman and Arnold [20] have also suggested that an imbalance between stroma and epithelium is at the core of carcinogenesis. They propose, however, the participation of stimulatory (growth factors, oncogenes) and inhibitory factors as signals that mediate what they call the 'balance point' of normality, that is, they subscribe to the notion that *quiescence* is the default state of metazoan cells. As mentioned in Chapter 8, the tissue organization field theory of carcinogenesis adopts the premise that *proliferation* is the default state of all cells.

The higher frequency of epithelial over connective tissue neoplasias is likely to reflect the physical connections among the cells in both tissues. In connective tissues, the highly organized intercellular space is larger and, presumably, provides more leeway for cell population expansion and for remedial adjustments. In epithelial tissues, cells are in close contact in a structured arrangement where numerical changes are usually handled by a wide array of options, from a minimal rate of apoptotic death to massive shedding into the lumen of the hollow organ. After local and organismal inhibitory mechanisms falter over a period of time, measured in cell population doubling times, a neoplasm may become palpable or visible.

These aberrant tissue patterns do not necessarily require the presence at their outset of genetically altered cells. Among the capabilities that may be expressed extemporaneously, motility is the most feared because these cells could move into other organs and recast a caricature of the tissue from which they derive (metastases).

In contrast to the somatic mutation theory, the tissue organization field theory allows for the phenomenon of regression. This may occur at any stage during the progression from the very initial stages of carcinogenesis to frank neoplasia. Since no DNA mutations are necessary, the neoplastic phenotype **is not** inexorably fixed. Experimental and clinical experience support these claims (see below).

## Experimental carcinogenesis: the lessons missed

Data are never theory-free. Specifically, experimental models in carcinogenesis and neoplasia have been selected for their consistency with, and their validation of, the somatic mutation theory. Models that questioned this theory were discarded, at least temporarily. We will briefly analyze data on foreign-body, physical, chemical, and viral carcinogenesis in the context of both the somatic mutation and the tissue organization field theories.

## Foreign-body carcinogenesis

It has been amply documented that foreign-bodies induce human cancers. However, the incidence of neoplasias generated by the implantation of medical devices (prostheses, steel mesh, pacemakers, plastic sponges, non-absorbable sutures, etc.) does not constitute a significant medical problem today. Nonetheless, it is well known that depending on the laboratory animal species adopted to conduct experiments, a different picture emerges. For instance, in guinea pigs foreign-bodies do not generate neoplasms at the site of implantation; instead, these materials readily produce tumors in individuals of susceptible mouse and rat strains [21]. Cellulose membrane filters of certain physical characteristics (implant size, shape, and porosity) generate neoplasias at the site of subcutaneous implantation. Histopathologically, the foreign-body generates an unspecific inflammatory response (neovascularization, white blood cell immigration, etc.), followed by a proliferative fibrotic reaction and subsequent transformation into a sarcoma capable of being sequentially transplanted into syngeneic hosts.

A comparable carcinogenic outcome follows exposure to special conformations of minerals in particulate form. Asbestos, beryllium, and other soft and hard materials are agents that may generate neoplasias in exposed individuals. The etiology of these neoplasias does not present problems for the pathologist as long as the source of the fibrotic reaction is identified in the histology slide, along with a history of exposure.

Remarkably, certain parasitic diseases are also associated with cancer. The most studied of these is bladder cancer associated with schistosomiasis (a disease caused by a parasite, *Schistosoma hematobium,* living in the rivers of several countries in Africa and Latin America) [22]. The common denominator between the foreign-body and the schistosomiasis-associated bladder cancers appears to be the chronic, local irritation in bladder venules and the stroma where parasite ova are lodged. The patient reacts with an initial nonspecific hyperplastic process and, later on, this normal tissue-repair reaction generates cells that escape the restraints imposed by the tissue organization field, and end up generating a full-blown local carcinoma.

What do all these foreign-body-associated cancers have in common? Size, shape, physical structure and texture of the agents are clearly related to the eventual emergence of the neoplastic disease. However, a direct interaction between the components of the foreign-bodies and the DNA is highly unlikely. Thus, there is no promising reductionistic approach to study foreign-body carcinogenesis, no clue in the host, nor in the noxious agents, that can be resolved into an experimental program that would shed light on the transformation of apparently 'normal' cells into neoplastic ones.

Little attention has been dedicated to foreign-body carcinogenesis, probably due to the low incidence of this type of neoplasia among populations of industrialized countries, and their 'bad' properties for a reductionistic approach. Supporters of the somatic mutation theory propose that would-be neoplastic cells in the host are already mutated in elusive oncogenes, and that the foreign-bodies play just a 'promotional' role in making the putative mutated cell become neoplastic. To the contrary, foreign-body carcinogenesis can be adequately explained by the tissue organization field theory of carcinogenesis. The host's reaction to the foreign-body represents another example of localized, altered tissue organization. No source, nor effect, of mutations in the constituent cells is required to explain it.

## Physical carcinogenesis

Fueled by generous funding after World War II, radiation biologists focused their research on uncovering details of how radiation generates neoplasias in humans and experimental animals. They also adopted a reductionist agenda, that is, they searched for the smoking gun in the DNA and in the DNA repair machinery mainly using cell culture models. The somatic mutation–excision/repair subordinate hypothesis used to explain carcinogenesis argues that radiation produces DNA breaks that, if misrepaired, will lead to DNA mutations. Xeroderma pigmentosa, a recessively inherited disease in which patients have ultraviolet light sensitivity, damaged DNA-repair mechanisms, and high incidence of skin neoplasias, was used as a showcase to boost this hypothesis [23]. These neoplasias were attributed to faulty DNA-repair mechanisms. However, published data raise relevant questions that challenge this hypothesis. Photosensitive tricothiodistrophy, a disease attributed to the same faulty repair enzyme present in xeroderma pigmentosa patients, *does not* result in an increased incidence of skin cancer. Remarkably, patients of both diseases have photosensitive skins and their cells *in culture* show hypermutability when exposed to ultraviolet light [24]. A comparable lack of fit has been described between the failure of DNA repair mechanisms and carcinogenesis in the Cockayne syndrome, another photosensitive condition [23]. A black box remains between the input, radiation, and the result, carcinogenesis. These conflicting data argue against the notion that there is a direct link between increased mutation rates and carcinogenesis. Much of the data collected using cells *in culture* were useful in clarifying what effects radiation had on cells as far as their viability and specific damage to subcellular structures. However, the integration of these data 'upward' to higher hierarchical levels has not fared as well [25].

In summary, the main conclusion that can be drawn from these data is that the failure of DNA repair mechanisms does not fully explain the increased carcinogenesis and neoplasia seen in exposed hosts. The link between mutations and carcinogenesis is, at best, circumstantial. Once again, what is at issue is not whether radiation increases the incidence of neoplasia, which is beyond dispute, but how and why this happens.

## Chemical carcinogenesis

In the 1940s, Peyton Rous [26] and Isaac Berenblum [27] proposed that skin neoplasias will develop following a two-step protocol that begins when cells that would eventually become neoplastic are permanently altered by a carcinogen ('initiation'). During the second step, the same 'initiated' cell will express its neoplastic phenotype when an ill-defined chemical interaction between a 'promoter' and its target cell takes place ('promotion'). The 'promotion' step is believed to involve cell death, increased cell proliferation (presumably resulting in more 'spontaneous' mutations), and clonal expansion of the more resistant cells (those that did not die from the action of the promoter). The two-step hypothesis assumed that the ultimate target for carcinogens was the cells of the epidermis.

It was in the 1970s that several researchers unambiguously stated that chemical carcinogens (*per se* or indirectly, through metabolic activation) acting as initiators, exerted their carcinogenic effects by inducing DNA damage and, hence, mutations [7, 9]. Demonstration of the mutagenicity of most of these chemicals was done using bacterial assays [28]. Promoters were not mutagenic in this model system. The use of bacteria was a pragmatic solution to the difficulty of evaluating mutagenesis in laboratory animals. Clearly, the definitions of initiators and promoters continue to be operational ones. Researchers working on carcinogenesis using target organs other than skin (mammary gland, liver, etc.) adopted the 'initiation/promotion' hypothesis in order to study their chosen models.

Data and interpretations challenging the above-mentioned inferences have been mostly disregarded. During the 1930s through to the 1950s, J.W. Orr produced evidence favoring, instead, the notion that the stroma under the epithelium of the skin was the crucially damaged tissue [18, 19]. Orr's data and conclusions, however, did not attract much attention. In fact, the cellular/subcellular target argument inherent in the somatic mutation theory won over that of Orr who explained carcinogenesis as a phenomenon happening at the tissue level of hierarchical complexity (see Chapter 7).

Many exceptions plagued the initiation/promotion hypothesis. For instance, Iversen showed that there was no clear-cut difference between the neoplasia-inducing effect of 'initiators' and 'promoters' on skin carcinogenesis [29]. With the passage of time, the original two-step initiation/promotion scheme was updated to fit new evidence and became a multistage process [30]. Proponents of the somatic mutation theory of carcinogenesis have readily acknowledged that, in some instances (as in the teratocarcinoma case, the foreign-body carcinogenesis, and others), epigenetic mechanisms may be sufficient to explain carcinogenesis.

Despite all the lacks of fit, the somatic mutation theory remains today's dogma in carcinogenesis. Ironically, in his acceptance speech for the 1966 Nobel prize, Peyton Rous said ' . . . A favorite explanation has been that (carcinogens) cause alterations in the genes of cells in the body, somatic mutations, as they are termed. But, numerous facts, when taken together decisively exclude this supposition' [31].

## Viral carcinogenesis. But . . . are viruses *per se* carcinogenic agents?

In the late 1960s, on the occasion of an anniversary of the creation of the Children's Cancer Research Foundation (the Jimmy Fund Building), in Boston, Massachusetts, James D. Watson explained to an attentive audience the inevitability that, *if* the DNA-containing Simian Virus 40 (SV40) induced neoplasias at the site of inoculation in rodents, it stood to reason that among the 5400 base pairs in its genome there had to be a gene responsible for those neoplasias. Again, the mystical power of the gene!

The overwhelming majority of human neoplasias were not then, nor are they now, considered to be associated with viruses, and for those neoplasias where viruses were suspected to play a causative role, the evidence favoring a **direct** link is becoming increasingly weaker. It remains unclear whether the viruses by themselves make *a* normal cell become *a* neoplastic one or, instead, viruses disrupt the ability of cells in tissues to communicate with one another. This latter option would favor the emergence of new social properties among the affected cells. This interpretation fits well with the tissue organization field theory of carcinogenesis. Supporting evidence for an **indirect** effect by viruses is suggested by the hepatitis B DNA virus, lacking putative oncogenes, involved in the pathogenesis of hepatocellular carcinomas in Africa and Asia [32]. The prevention of this scourge has been achieved by an effective vaccine against the virus, thus averting hepatitis and the resulting damage to the liver. Most interestingly, both viral and non-viral hepatitis increase the risk of hepatocellular carcinomas; the common denominator in both cases is an inflammatory process which often leads to a fibrotic reaction (cirrhosis) and then to carcinoma.

The data accumulated on viral carcinogenesis have allowed for multiple interpretations of the roles viruses play in this complex process. One of them spun off into the *oncogene hypothesis*, a subordinate hypothesis within the somatic mutation theory. The oncogene hypothesis, also based on the premise that the default state in metazoa is *quiescence*, proposed that the 'carcinogenic viruses' were the Trojan horses that carried the true cancer genes. Oncogenes were the mutated versions of normal cellular genes carried by those RNA viruses. The cellular oncogenes were thought to be either cell cycle effectors, receptors for growth factors, mediators of growth factor action in the signal transduction pathway, or transcription factors [33]. It was assumed that these mutated oncogenes were 'activated' constitutively, and hence, pushed the host cells first into proliferation and subsequently into neoplasia due to the clonal expansion of a single affected cell. Interestingly, the viral oncogenes do not play a significant role in the survival or replication of the viruses of which they are a part [34].

DNA viruses were also claimed to **directly** generate neoplasias [33]. Once integrated into the host cell DNA, they would affect the cell cycle by forcing the cell to synthesize its own DNA, repeatedly inducing its proliferation, and as a result generate a tumor. The genes in these DNA viruses responsible for the activation of the host DNA replication are not homologous to the host's own genes. Hence, they are different from the retroviral oncogenes. A hypothesis on the mechanism for DNA viral carcinogenesis had to await the discovery of so-called anti-oncogenes or suppressor genes.

Epidemiological data on hereditary cancers were compatible with the existence of genes that when 'inactivated', or missing, resulted in the appearance of neoplasias [35]. The current interpretation of how DNA viruses may induce neoplasias evolved from the discovery of the retinoblastoma gene. Hereditary retinoblastoma is an infrequent tumor that appears in the eyes of young children whose parents carry one damaged copy of a gene that was thus called the Rb gene. According to this hypothesis (the *two-hit hypothesis*), the normal allele becomes mutated only in the tumors appearing in the eyes of those children [33]. It was postulated that the Rb protein and other anti-oncogene products acted as cell cycle inhibitory factors. Oncogene supporters accommodated the new arrivals when it was shown that viral DNA gene products, such as SV40 large T antigen, adenovirus E1A, and papilloma viruses E6 and E7, bound to this suppressor gene products. In this way, putative inhibitors of cell cycle progression would become neutralized by these viral gene products, and the host cells would be sent on their way to proliferate and form the neoplasia. Thus, a new 'guilt by association' argument has been advanced, whereby these viral gene products would abolish the function of cell cycle inhibitors.

### *Is there room for a compromise (a synthesis) between competing hypotheses?*

Would it be possible to integrate data generated while adopting the somatic mutation theory into the tissue organization field theory of carcinogenesis? Specifically, how do mutations in suppressor genes, like Rb and p53, fit into the tissue organization field theory? As mentioned in Chapters 1, 5 and 6, acknowledging that biological complexity is not the work of an engineer but that of a tinkerer, the blueprint for this complexity is unavailable to researchers. Thus, we explore theories to verify how well data fit our preconceived notions. In other words, we attempt to reach good approximations to a complex reality. Rb, p53, and other putative suppressor genes may or may not end up fitting our preferred theory. For the time being, the data stemming from gene inactivation experiments (knockouts) suggest an inadequate fit even within the somatic mutation theory [15]. In this book we are not addressing the question 'Which is the role of these suppressor genes?'; instead, the question that has occupied us is, 'How does carcinogenesis evolve?' or 'At what hierarchical level of complexity does carcinogenesis occur?' Hence, we are challenging the notion that somatic mutations generate neoplasia. This does not mean that we question that mutations do occur in somatic cells. We even acknowledge that the disruption of tissue and individual cell architecture in tumors may increase the rate of chromosomal aberrations, and even mutations in those cells. By chance, those mutations may affect gene products involved in the communication among cells and hence, they would increase the possibilities that those cells may escape natural inhibitory signals of cell proliferation and motility. These would obviously aggravate the prognosis of these neoplasias. However, these constitute secondary phenomena most likely unrelated to the original lesion occurring at the tissue organization level responsible for the initiation of the carcinogenic phenomenon.

J.M. Bishop, a 1989 Nobel prize awardee for the discovery of oncogenes, boldly stated ' . . . we have learned very little about the causes of cancer from the study of proto-oncogenes and tumor suppressor genes' [36]. However, in Bishop's words, there is a 'compelling logic' to the finding that oncogenes code for growth factor receptors, proteins involved in signal transduction, transcription factors, and cell cycle effectors [37]. Hence, a circular argument is built whereby the significance of a growth factor stems from its association with an oncogene, and the relevance of an oncogene resides in its association with a growth factor. The flip-side of the argument, namely, what type of association **would not** have been indicative of a 'compelling logic' for the relevance of oncogenes to neoplasia, is worth considering. For example, if oncogenes coded for mitochondrial enzymes involved in respiration, would this be considered trivial or supportive of Warburg's hypothesis [38]? If oncogenes coded for proteins that attach cells to the basement membranes or to other cells, would it disqualify oncogenes, or would it make a compelling argument for autonomy of the inhibitory influences of the tissue organization field? Finally, we found a few instances whereby possible associations were outright preposterous. One example of this would be an oncogene coding for hemoglobin, the oxygen-carrying protein of red blood cells; this is a gene product expressed in cells that are committed to lose their nuclei and die. These considerations plainly demonstrate how difficult it would be to vindicate or falsify the latter day version of the somatic mutation theory. In summary, it is only an idiosyncratic 'compelling logic' that makes researchers consider mutations, oncogenes and anti-oncogenes as directly responsible for carcinogenesis.

## On the uniqueness of the neoplastic cell

### *Neoplastic cells that are not per se neoplastic?*

The implicit message of the somatic mutation theory of carcinogenesis is 'Once a cancer cell, always a cancer cell' [39]. Does the evidence support this notion? This question was addressed using several experimental models. Teratocarcinomas are tumors that arise spontaneously in the testis of certain mouse strains. These tumors look like a caricature of a normal embryo; many normal-looking tissues and cell types are present in them. Almost four decades ago, it was shown that mouse teratocarcinoma cells could behave as multi-potential normal cells [40]. Teratocarcinoma cells were found to act as stem cells, generating all the recognizable differentiated tissues. When a single cell of these different cell types was transplanted into mice, only the teratocarcinoma stem cells gave rise to neoplasias. The differentiated cells derived from these stem cells did not form tumors. Thus, genuine neoplastic cells were able to generate normal cells and normal tissues. These findings contradict the concept that the neoplastic, malignant behavior is 'fixed'. Now, if neoplasia, as posited by the somatic mutation theory, is due to the accumulation of multiple mutations, how can cells derived from these neoplasias revert to behave as normal cells? It is statistically unlikely that random reverse mutational events that could erase the previous mutations would be responsible for this reversal.

Almost 25 years ago, Illmensee and Mintz showed that a single teratocarcinoma cell was able to generate normal tissues and organs when transplanted into early normal embryos (blastocysts) [41]. These neoplastic cells proliferated in suspension in the peritoneal cavity of mice and had been continuously transplanted for over 8 years from mouse to mouse (about 200 transplants) previous to them being inoculated into the blastocysts. These blastocysts developed into viable, healthy mice that were, in fact, mosaics of tissues derived from a mixture of normal embryonal cells from their 'natural' parents, along with a pool of cells derived from the single transplanted neoplastic cell. From these mosaic mice, normal progeny were generated by sperm derived from cells that were once teratocarcinoma cells. These data are consistent with the inference made above that *a* single neoplastic cell may, in fact, be a normal cell as far as its proliferative capability and its ability to express the genomic blueprint it carries. This is accomplished by responding appropriately to cues from the surrounding cells and extracellular matrix.

The road from normalcy to neoplasia is not unidirectional. Cells from a mammalian embryo may generate neoplasms when placed outside their normal habitat (under the kidney or testis capsule, or in the peritoneum). Once these neoplastic cells are placed back into a 'friendly' embryonal environment, they may reacquire a normal phenotype and proliferative behavior. These properties are determined by the cells' position in their adult tissue organization field (or in the morphogenetic field in the embryo). Taken together, data from these experimental models support the notion that neoplasias are emergent phenomena happening at the tissue level. Equally evident, the somatic mutation theory cannot satisfactorily account for these data.

### *The meaning of aneuploidy in carcinogenesis and neoplasia*

A long-dated controversy has raged about the role played by 'ploidy' and 'differentiation' in the neoplastic cells' ability to proliferate. Ploidy refers to the variations in number and structure in the set of chromosomes carried by each neoplastic cell. Boveri's original proposal that the gain or loss of chromosomes during mitosis was at the center of carcinogenesis could not be exhaustively explored until the late 1950s, when the normal human karyotype, a display of 22 pairs of somatic chromosomes and one pair of sex chromosomes, was described [42], and its detailed study was made easier in the 1960s by technical improvements, such as chromosome banding which allowed for their more detailed identification. A number of neoplasias were shown to have aneuploid cells. However, neoplastic cells also showed karyotypes not different from normal ones. In the early 1970s, it was noticed that in chronic myelocytic leukemias a majority of cells showed a fairly constant chromosomal anomaly, that is, the deletion of the long arm of chromosome 22 which often became translocated into the long arm of chromosome 9 [43]. Technical difficulties have impaired the study of the chromosomal pattern of neoplastic cells in solid tumors, which represents the vast majority of neoplasias, mainly of epithelial origin. This has prevented the drawing of reliable conclusions over the prevalence of non-random chromosomal markers in these neoplasias. The presence of specific, non-random chromosomal markers was interpreted by some as being causal, and consequently, a manifestation of the monoclonality (single-cell origin) of neoplasias. However, monoclonality may result from other pathways. For example, a selective process may operate on a heterogeneous cell population in which a cell manages to reproduce faster than its neighbors and ends up having more progeny; this would drive cells that did not proliferate so efficiently to undetectable numbers [43, 44].

It is impossible today, as it was in Boveri's time, to observe a neoplasia 'in statu nascendi', that is, as it originates [2]. Thus, the same set of data may be explained differently depending on the *a priori* assumptions that we adopt. If one subscribes to the somatic mutation theory, these unique chromosome markers spell monoclonality, and are likely to be causal. From the tissue organization field theory viewpoint, the same marker represents an epiphenomenon.

One indirect way to gain further knowledge on the relevance of these chromosomal markers and differentiation is to test experimentally the reversibility of the neoplastic behavior of cells when placed in normal environments in susceptible strains of animals. Grisham's group has contributed to this goal [45]. Briefly, rat liver cells became 'established' *in culture* following a treatment with chemical carcinogens. When these cell lines were injected subcutaneously into inbred rats, 'undifferentiated' (spindle-shaped cell) neoplasias developed at the site of inoculation, regardless of whether the cells carried a euploid (normal) or aneuploid complement of chromosomes. Alternatively, when these cell lines were injected now inside the liver, only some of them developed hepatoma-like tumors. Remarkably, cells from the cell lines that did not develop hepatomas became normally integrated into the liver parenchyma regardless of their ploidy. These data strongly support the notions that, first, aneuploidy was not an impediment for cells to become 'normalized' when placed in a normal liver microenvironment and, second, the phenomena described by Illmensee and Mintz are not limited to a single, diploid cell closely resembling a normal embryo cell since Grisham's experiments involved the inoculation of euploid *and* aneuploid neoplastic liver cells. Thus, neoplastic cells other than totipotential teratocarcinoma cells may also become 'normalized' when exposed to a permissive microenvironment. Finally, regardless of whether or not all these cells were mutated in a yet-to-be-identified key gene, the data suggest that these mutations had little effect in perpetuating the neoplastic phenotype.

## Carcinogenesis in a flask

How do data gathered *in culture* fit within carcinogenesis where it really happens, that is, in animals, including humans? As mentioned earlier, data are not theory-free. For those who adopt the somatic mutation theory to study carcinogenesis, cellular/subcellular events conducted *in culture* provide a legitimate experimental device, since histoarchitecture (tissue organization) does not play a central role in this theory. To the contrary, from the tissue organization field theory perspective, there are two alternative ways to deal with *in-culture* data stemming from the cellular/subcellular hierarchical level. The first is to dismiss outright the interpretation of the data as irrelevant to carcinogenesis because cell culture experiments are unsuitable to describe the initial stages of carcinogenesis seen in animals. The second is to assume that the cells 'transformed' *in culture* are equivalent to frank neoplastic cells, which arose from normal cells through a process that does not recapitulate carcinogenesis. 'Demergence' *in culture* removes the constraints imposed by the organism on these cells, and may in some cases result in cells that, when placed into syngeneic hosts, would grow as tumors [46–50].

From the tissue organization field theory perspective, carcinogenesis results from the disruption of cell-to-cell communication involving the parenchyma and its stroma. Thus, it follows that to study carcinogenesis using *in-culture* models, it is necessary to preserve **tissue** organization. This implies that tissue and organ culture would be the appropriate models to study carcinogenesis. However, the use of these models is limited in practice today, due to the short lifespan of explants. Nonetheless, clear goals promote effective solutions. Although cell culture began to be used at the dawn of the current century, its impact in biology started towards the end of the 1960s when researchers and suppliers recognized the need to significantly improve and standardize the quality of the relevant materials and equipment (serum, plasticware, laminar hoods, etc.). The tissue and organ culture models we envisage can be developed and used advantageously to study carcinogenesis under solid biological principles. This includes the transplantation of these tissues back into animals, and the verification that the observations recorded from *in-culture* experiments may be integrated with those observed when using animal models. For all practical purposes, 'carcinogenesis in a flask', a quintessential cellular/subcellular model, has not significantly helped in the understanding of carcinogenesis where it really happens, that is, in a tissue and in an organism.

## Conclusions

During the current century, and especially during the last three decades, the somatic mutation theory to explain carcinogenesis became dogma. The use of powerful tools and approaches provided by the molecular biology revolution has permitted a thorough exploration of this theory. However, no answer has been forthcoming on whether or not genes are the ultimate targets of carcinogens. It is still unknown which genes had to be mutated in order to trigger the long-term ability of those damaged cells to proliferate extemporaneously. Equally elusive has been an explanation emanating from this theory that would integrate mutations with the abnormal tissue organization typical of neoplasia. Instead, contradictory evidence has been incorporated into the body of the theory by providing *ad hoc* explanations. For example, in foreign-body carcinogenesis, which cannot be explained through mutations, foreign-bodies are considered to act as 'promoters'. That is, it is assumed that mutations were already present in cells at the site of implantation, and the proliferative activity elicited by the foreign-body generated more mutations and amplification of the pool of mutated cells. When the hypothetical chromosomal lesions in neoplastic cells could not easily account for the monoclonality of tumors, point mutations and lesser structural aberrations were invoked as the *primum movens* in neoplastic development. As room is made to reconcile every single improper fit, there is no chance for falsifying any hypothesis. Thus, nothing can be ruled out. This situation contrasts with the objectives of science as described by Ayala [51]: (1) science seeks to organize knowledge in a systematic way, endeavoring to describe patterns of relationship between phenomena and processes; (2) science strives to provide explanations for the occurrence of events; and finally, and pointedly relevant to the topic of theories on carcinogenesis, (3) science proposes explanatory hypotheses that must be testable, that is, accessible to the possibility of rejection.

The tissue organization field theory of carcinogenesis adopts the premise that *proliferation* is the default state of all cells. It further proposes that the stroma and parenchyma are the targets of changes elicited by carcinogens. This theory lends itself to be studied using the morphogenetic field paradigm. During development, the temporal and spatial expression of genes governs the developmental fate of cells. Hence, cells 'know' where they come from (historical information), where they are (positional information), and this information limits their fate to a restricted phenotypic choice. During adulthood, interactions between stroma and parenchyma continue to regulate tissue maintenance and regeneration. From this perspective, the exploration of the tissue organization field theory of carcinogenesis requires the use of tissue recombination and transplantation experiments, comparable to the ones that informed embryologists about the inductive and permissive influences in organogenesis.

The fruit fly *Drosophila* offers a great opportunity to study carcinogenesis from the perspective of the morphogenetic field paradigm using a genetic approach. The genome of this animal is almost completely sequenced, and an awesome collection of mutants is available. Hence, carcinogenesis, as described in the *lethal giant larvae-2* and other mutants, could be followed chronologically by comparing the pattern of gene expression in the organ system where the neoplasia will appear in mutated larvae with the pattern occurring in *wild-type* specimens of the same age. In vertebrates, this level of complexity may be explored by establishing whether a given carcinogen that induces neoplasias in a specific organ acts primarily on the stroma or the parenchyma; this can be accomplished by following the fate of recombinations between carcinogen-exposed tissues and untreated ones. The now flourishing field of developmental genetics appears ready to provide conceptual and methodological support to the study of organizational disruptions at the tissue level in both vertebrates as well as invertebrates. We predict that this is the correct direction for the field of carcinogenesis to follow.

## References


1. **Mayr, E.** (1996) *This is Biology.* Harvard University Press, Cambridge, MA.
2. **Boveri, T.** (1929) *The Origin of Malignant Tumors.* Williams & Wilkins, Baltimore, MD, p. 115.
3. **Avery, O.T., MacLeod, C.M. and McCarty, M.** (1944) Studies on the chemical nature of the substance inducing transformation of pneumococcal types. *J. Exp. Med.* **79**: 137–158.
4. **Watson, J.D. and Crick, F.H.C.** (1953) Molecular structure of nucleic acids: a structure for deoxyribose nucleic acid. *Nature* **171**: 737–738.
5. **Gilbert, S.F., Opitz, J.M. and Raff, R.A.** (1996) Resynthesizing evolutionary and developmental biology. *Dev. Biol.* **137**: 357–372.
6. **Gurdon, J.B.** (1968) Transplanted nuclei and cell differentiation. *Sci. American* **219**: 24–35.
7. **Ames, B.N., Durston, W.E., Yamasaki, E. and Lee, F.D.** (1973) Carcinogens as mutagens: a simple test system combining liver homogenates for activation and bacteria for detection. *Proc. Natl Acad. Sci. USA* **70**: 2281–2285.
8. **Miller, J.A., Cramer, J.W. and Miller, E.C.** (1960) The n- and ring-hydroxilation of 2–acetylaminofluorene during carcinogenesis in the rat. *Cancer Res.* **20**: 950–962.
9. **Miller, J.A. and Miller, E.C.** (1977) Ultimate chemical carcinogens as reactive mutagenic electrophiles. In: *Origins of Human Cancer* (eds H.H. Hiatt, J.D. Waston and J.A. Winsten). Cold Spring Harbor Laboratory Press, Plainview, NY, pp. 605–627.
10. **Alberts, B., Bray, D., Lewis, J.G., Raff, M., Roberts, K. and Watson, J.D.** (1994) *Molecular Biology of the Cell*, 3rd Edn. Garland Publishing, New York, pp. 863–910.
11. **Hunt, T. and Nasmyth, K.** (1997) Cell multiplication. *Curr. Opinion Cell Biol.* **9**: 765–767.
12. **Duronio, R.J., O'Farrell, P.H., Xie, J-E., Brook, A. and Dyson, N.** (1995) The transcription factor E2F is required for S phase Drosophila embryogenesis. *Genes Develop.* **9**: 1445–1455.

13. **Herrera, R.E., Sah, V.P., Williams, B.O., Makela, T.P., Weinberg, R.A. and Jacks, T.** (1996) Altered cell cycle kinetics, gene expression, and G1 restriction point regulation in Rb-deficient fibroblasts. *Mol. Cell. Biol.* **16**: 2402–2407.

14. **Nakayama, K., Ishida, N., Shirane, M., Inomata, A., Inoue, T., Shishido, N., Horii, I. and Loh, D.Y.** (1996) Mice lacking p27 (Kip1) display increased body size, multiple organ hyperplasia, retinal dysplasia, and pituitary tumors. *Cell* **85**: 707–720.

15. **Wang, J.Y.J.** (1997) Retinoblastoma protein in growth suppression and death suppression. *Curr. Opinion Genet. Develop.* **7**: 39–45.

16. **Ohtani, K., DeGregori, J. and Nevins, J.R.** (1995) Regulation of the cyclin E gene by transcription factor E2F1. *Proc. Natl Acad. Sci. USA* **92**: 12146–12150.

17. **Farber, E.** (1984) Chemical carcinogenesis: a current biological perspective. *Carcinogenesis* **5**: 1–5.

18. **Orr, J.W.** (1955) The early effects of 9:10-dimethyl-1:2-benzanthracene on mouse skin, and their significance in relation to the mechanism of chemical carcinogenesis. *Br. J. Cancer* **9**: 623–632.

19. **Orr, J.W.** (1958) The mechanism of chemical carcinogensis. *Br. Med. Bull.* **14**: 99–101.

20. **Kaufman, D.G. and Arnold, J.T.** (1996) Stromal–epithelial interactions in the normal and neoplastic development. In: *Cell and Molecular Pathogenesis* (ed. A.E. Sirica). Lippincott-Raven, Philadelphia, PA, pp. 403–432.

21. **Brand, K.G.** (1982) Cancer associated with asbestosis, schistosomiasis, foreign bodies, and scars. In: *Cancer: A Comprehensive Treatise* (ed. F.F. Becker). Plenum Press, New York, pp. 661–692.

22. **Elsebai, I.** (1977) Parasites in the etiology of cancer—Bilharziasis and bladder cancer. *Ca: Cancer J. Clinicians* **27**: 100–106.

23. **Kern, S.E.** (1996) Oncogenes/proto-oncogenes, tumor-supressor genes, and DNA-repair genes in human neoplasia. In: *Cell and Molecular Pathogenesis* (ed. A.E. Sirica). Lippincott-Raven, Philadelphia, PA, pp. 321–340.

24. **Bridges, B.A.** (1981) Some DNA-repair-deficient human syndromes and their implications for human health. *Proc. Roy. Soc. B.* **212**: 263–278.

25. **Little, J.B.** (1994) Changing views of cellular radiosensitivity. *Radiation Res.* **140**: 299–311.

26. **Friedewald, W.F. and Rous, P.** (1944) The initiation and promoting elements

in tumor production. An analysis of the effects of tar, benzpyrene, and methylcholantrene on rabbit skin. *J. Exp. Med.* **80**: 101–130.

27. **Berenblum, I. and Shubik, P.** (1947) A new quantative approach to the study of the stages of chemical carcinogenesis in the mouse's skin. *Br. J. Cancer* **1346**: 383–391.

28. **Ames, B.N.** (1979) Identifying environmental chemicals causing mutations and cancer. *Science* **204**: 587–593.

29. **Iversen, O.H.** (1993) Role of cell proliferation in carcinogenesis: is increased cell proliferation in itself a carcinogenic hazard? In: *New Frontiers in Cancer Causation* (ed. O.H. Iversen). Taylor & Francis, Washington, DC, pp. 97–105.

30. **Sirica, A.E.** (1996) Multistage carcinogenesis. In: *Cellular and Molecular Pathogenesis* (ed. A.E. Sirica). Lippincott-Raven, Philadelphia, PA, pp. 283–320.

31. **Rous, P.** (1966) The challenge to man of the neoplastic cell. *Les Prix Nobel.* Imprimerie Royale, Stockholm, pp. 162–171.

32. **Payne, R.J., Nowak, M.A. and Blumberg, B.S.** (1992) Analysis of a cellular model to account for the natural history of infection by the hepatitis B virus and its role in the development of primary hepatocellular carcinoma. *J. Theor. Biol.* **159**: 215–240.

33. **Alberts, B., Bray, D., Lewis, J.G., Raff, M., Roberts, K. and Watson, J.D.** (1994) *Molecular Biology of the Cell*, 3rd Edn. Garland Publishing, New York.

34. **Isom, H.S., Wigdahl, B. and Howett, M.K.** (1996) Molecular pathology of human oncogenic viruses. In: *Cellular and Molecular Pathogenesis* (ed. A.E. Sirica). Lippincott-Raven, Philadelphia, PA, pp. 341–387.

35. **Knudson, A.G., Jr** (1995) Mutation and cancer: a personal odyssey. *Adv. Cancer Res.* **67**: 1–23.

36. **Bishop, J.M.** (1991) Molecular themes in oncogenesis. *Cell* **64**: 235–248.

37. **Bishop, J.M.** (1987) The molecular genetics of cancer. *Science* **235**: 305–311.

38. **Warburg, O.** (1956) On the origin of cancer cells. *Science* **123**: 309–314.

39. **Pierce, G.B., Shikes, R. and Fink, L.M.** (1978) *Cancer: A Problem of Developmental Biology.* Prentice-Hall, Englewoods Cliffs, NJ.

40. **Pierce, G.B., Dixon, F.J. and Yerney, E.L.** (1960) Terato-carcinogenic and tissue forming potentials of the cell types comprising neoplastic embryoid bodies. *Lab. Invest.* **9**: 583–602.

41. **Illmensee, K. and Mintz, B.** (1976) Totipotency and normal differentiation of single teratocarcinoma cell cloned by injection into blastocysts. *Proc. Natl Acad. Sci. USA* **73**: 549–553.

42. **Tjio, J.H. and Levan, A.** (1956) The chromosome number of man. *Hereditas* **42**: 1–6.

43. **Rowley, J.D. and Mitelman, F.** (1993) Principles of molecular cell biology of cancer: chromosome abnormalities in human cancer and leukemia. In: *Cancer: Principles & Practice of Oncology* (eds V.T. DeVita, S. Hellman and S.A. Rosenberg). J.B. Lippincott Company, Philadelphia, PA, pp. 67–91.

44. **Heim, S. and Mitelman, F.** (1987) *Cancer Cytogenetics*. A.R. Liss, New York.

45. **Coleman, W., Wennerberg, A.E., Smith, G.J. and Grisham, J.W.** (1997) Regulation of the differentiation of diploid and aneuploid rat liver epthelial (stem-like) cells by the liver microenvironment. *Am. J. Pathol.* **142**: 1373–1382.

46. **Earle, W.R.** (1943) Production of malignancy in vitro. IV. The mouse fibroblast cultures and changes seen in the living cell. *J. Natl Cancer Inst.* **4**: 165–212.

47. **Gey, G.O.** (1955) Some aspects of the constitution and behavior of normal and malignant cells maintained in continuous culture. *Harvey Lect.* **50**: 154–229.

48. **Sanford, K.K.** (1965) An attempt to induce malignant transformation of cells in vitro by intermittent anaerobiosis. *J. Natl Cancer Inst.* **35**: 719–725.

49. **Paquette, B., Wagner, R. and Little, J.B.** (1996) In vitro reversion of transformed phenotype in mouse C3H-10T1/2 cells: modification in genomic 5-methylcytosine content. *Int. J. Oncol.* **8**: 727–734.

50. **Rubin, H.** (1985) Cancer as a dynamic developmental disorder. *Cancer Res.* **45**: 2935–2942.

51. **Ayala, F.J.** (1968) Biology as an autonomous science. *Am. Scientist* **56**: 207–221.

Epilogue

# Moving Toward the Integration of Cell Proliferation, Carcinogenesis, and Neoplasia into Biology

"An astonishingly high proportion of major new concepts and theories is based on components that had long before been available but which no one has been able to tie together properly."

*Ernst Mayr (1982) The Growth of Biological Thought. Harvard University Press, Cambridge, MA, p. 852*

"Most serious of all the results of the somatic mutation hypothesis has been its effects on research workers. It acts as a tranquilizer on those who believe in it, and this at a time when every worker should feel goaded now and again by his ignorance of what cancer is."

*Peyton Rous (1959) Surmise and fact on the nature of cancer. Nature* ***183****: 1357–1361*

This is a book about concepts, data, and interpretations. Throughout its chapters, we have contended that the understanding of the control of cell proliferation and cancer has been hindered by unstated ideologies and operational definitions. We hope that our analysis has persuaded you, the reader, of the importance for unveiling hidden premises chosen by researchers when designing experiments and interpreting data. Because premises cannot be avoided, they must be acknowledged and dealt with up front.

# Ideology, silent assumptions, and operational definitions

Steven Weinberg, the physicist, stated that the usefulness of the philosophy of science to scientists is comparable to the usefulness of the aerodynamic theory to birds. We cannot argue about whether or not this attitude is advantageous to the search for knowledge in the physical sciences. In the biological sciences, this disregard for philosophy is widespread. We are convinced, however, that this attitude has resulted in major problems, since experimental design is influenced by ideology and epistemology. Assumptions about which concepts are accepted as 'true', and which are rejected as 'false' are seldom articulated by biologists.

## *Ideology*

In many colleges and universities in the USA, departments of biology are literally being split in two: departments of Integrative and Evolutionary Biology and of Molecular Biosciences. Why is this happening? The success of reductionism in clarifying many important issues in biology (e.g. the understanding of how genetic information is coded and transmitted, and the phenomenon of gene control by environmental cues) emboldened a number of its practitioners to proclaim that *all* phenomena of life could be reduced to genes and their expression. This ideology dispenses with the hierarchical organization of life, and it produces the false assurance that all biological questions may be answered at the molecular level of inquiry. To the believers of this ideology, biology is reduced to molecular biology, and the gene becomes the unit of everything, selection included. Hence, intelligence, gender preference, violent or kind behavior, cancer, and most diseases are construed to be Mendelian traits. For the last quarter of a century, François Jacob, a pioneer of the molecular biology revolution, wisely warned against this attitude.

The 200-year-old dispute between reductionists and organicists seems to have transformed itself into a total rift. John Tyler Bonner stated, 'What is utterly baffling to me is why one cannot be a reductionist and a holist at the same time.' An answer to this inquiry is that the rift is ideological, rather than methodological. The rift originates, on the one hand, in the refusal by reductionists to acknowledge the principle of the hierarchical organization of nature, and thus, of emergence. Strict reductionism is deterministic, linear, non-hierarchical, and proposes that all biological questions should be answered at the molecular level. On the other hand, limited research is currently performed on the control of cell proliferation and carcinogenesis from an organicist perspective. In part, this is due to the prevalence of reductionist thinking among those who decide which type of research is funded, and admittedly, to the paucity of research projects reflecting the organicist perspective. Developmental biologists, for their part, have taken an eclectic approach whereby an organicist rationale takes advantage of the tools of reductionism [1]. This instance suggests that Bonner's conundrum may still be resolved. Ultimately, as Dobzhansky remarked, 'Nothing in Biology makes sense except in the light of evolution.'

According to Ernst Mayr, there are three types of biological questions: what, how, and why. The answers to 'what' questions are descriptive; science cannot be done without a solid description of the findings in which a theory is based. In this century, the word 'descriptive' has taken a pejorative connotation, and is used by some scientists to belittle the effort of colleagues (my research is mechanistic, yours is descriptive). Descriptive research applies to all levels of complexity, despite protestations by those who, working at the molecular level, think otherwise. 'How' questions address the function of the organism at all hierarchical levels, and 'why' questions address evolutionary phenomena [2]. All biological phenomena are subject to this analysis.

### *Operational definitions*

Biologists deal with operational definitions out of necessity, since we are piercing into a black box that we have not designed, and therefore, we can hardly second guess. When exploring a given phenomenon, there is a lot more than meets the eye. However, anything that goes into the black box and produces an effect is anthropocentrically interpreted as a 'stimulus', an 'inducer' or a 'positive regulator'. If, instead, what goes into the black box prevents the occurrence of a phenomenon that would have taken place without our intervention, the modifier becomes a 'blocker', a 'repressor', or a 'negative regulator'. This usage is acceptable as long as we recognize that those are operational definitions, which by no means suggest that our intervention is any more than a proximate cause. Suppose, for instance, that the 'stimulus' was EPO. The observed effect of this hormone is an increase in the red blood cell count. The operational definition of EPO as being a positive regulator (growth factor), becomes meaningless once we know the details of the process by which it increases erythrocyte numbers, namely, by inhibiting the death of erythrocyte progenitors. In spite of these data, EPO is still generally considered a growth factor. In conclusion, the processes by which growth factors and oncogenes sprung as preeminent characters in the fields of control of cell proliferation and cancer represent textbook examples of the mistaken rationale that started with harmless operational definitions.

## The universality of *proliferation* as the default state of all cells

Approximately 4 billion years ago prokaryotes appeared on Earth. About 900 million years ago, the first eukaryotes evolved from the prokaryotes. This generation of eukaryotes from prokaryotes involved the conservation of the pre-existing levels of organization and the emergence of higher ones. What happened to these previously free-living beings that caused them to become parts of new, more complex organisms? Why did they relinquish their autonomy? Why did the new arrangements become fixed? At the beginning of these symbiotic arrangements, the old prokaryotes which were incorporated into the new life forms must have gained versatility while still retaining the ability to propagate as a part of the new organism [4]. Conflicts may have emerged among intracellular self-replicating organelles and also between them and their host. If the rate of proliferation of any one organelle (mitochondria, for instance) impaired the survival of the host cell, the organism would have lost its chance to propagate. Hence, the evolution of these eukaryotes must have involved the loss of units of selection. For example, a portion of the organelle's DNA became incorporated into nuclear DNA, and thus, it became interdependent with that of the latter.

The first multicellular organisms appeared about 200 million years after the first eukaryotes evolved on Earth. It is acknowledged that for the last 600 million years, no new successful body plans have evolved in metazoa, while plants have generated new ones since that time. The reason for the rapid fixation of body plans in metazoa is not apparent; it is tempting however, to speculate that the segregation of the germ line in metazoa may have imposed the constraints to limit the body plan options.

The generation of multicellular organisms from unicellular eukaryotes also involved the conservation of pre-existing levels of organization. The built-in capacity for self-replication of cells within a multicellular organism must have remained unaltered. Evidence arguing for this conclusion is based on the fact that multicellular organisms develop from a single cell (the zygote) that in many species initiates development outside the parental organism.

The finding of almost complete homology between the machinery to replicate yeast and human cells suggests that the process of cell replication in eukaryotes has remained constant throughout evolution. Like the organisms from which they evolved, cells in multicellular organisms multiply as long as nutrients are available. What has changed and become the core of the selection process must have been the ability of the cells in the multicellular organisms to express different phenotypes as dictated by the developmental program [3]. As organisms develop from a single cell, the materialization of a body plan implies the cooperation of cells in the phenomena of induction, and the coordination of the proliferative activity of each lineage; determinants placed in the egg by the mother assure an orderly beginning.

Metazoa, like metaphyta, retained the properties of their single cell ancestors, but evolved emergent properties such as cell-to-cell communication, and 'reverse emergence' properties, in which the organism imposes controls over its parts. In multicellular organisms, there are quiescent cells. However, there is no reason to believe that this quiescent state is a newly acquired default state, rather than the consequence of a regulatory event imposed on specific cell types. The prevailing idea that *quiescence* appeared as a new default state in metazoa has never been adequately supported by either argument or data. Nevertheless, it has been accepted as dogma until now.

How can organisms relinquish the fundamental property of self-replication? If, for the sake of argument, one is determined to accept *quiescence* as a newly acquired default state in metazoa, conceptual problems emerge because the germ line in these animals should obligatorily maintain the ability to self-replicate, a truly *sine qua non* of life. If one proposes that only the somatic cells lost this property, those who propose this novelty have to explain how, when, and why this switch has taken place in the course of evolution. Two arguments favor the maintenance of *proliferation* as the default state for all cells. First, the genome is similar in all cells of a metazoan, and in experimental conditions, it has been shown that somatic cell nuclei may generate whole individuals when placed in enucleated oocytes. Second, the segregated germ cells, like their somatic counterparts, are subject to control of cell proliferation, as shown by the 'dormancy' of oogenesis and spermatogenesis at certain developmental stages. If the built-in capacity of these cells to proliferate was not curtailed by organismal control, their exponential proliferation would have destroyed the soma.

In short, our proposition could be fairly construed as an effort to put the default state of all cells back into a conceptual *status quo ante*. What has emerged as novelties in metazoa are signals which stop cell proliferation when the supply of nutrients would have allowed the cells to proliferate. It remains puzzling, however, why no overt attempt was mounted to challenge the anthropocentric premise of *quiescence* as the default state for metazoa. This re-interpretation of data on a topic so central to the concept of life has not just an academic interest. From a more pragmatic, medical viewpoint, the implications on the etiopathogenesis of cancer cannot be overemphasized.

## Forerunners of the tissue organization field hypothesis

Needham [5], Waddington [6] and Orr [7] proposed that neoplastic development resulted from a failure of the normal interactions that occur in morphogenetic fields. This idea was abandoned due to several factors. One reason was the ascending prestige of genetics, which was predicated by Morgan and his followers using rhetoric aimed at discrediting morphogenetic fields as mystic concepts [8]. Another reason was the continuing popularity of the somatic mutation theory throughout this century. Those factors prevented a serious consideration of morphogenetic fields as the substrate of carcinogenesis. The final factor for abandoning this theory was the experimental data interpreted as evidence that morphogenetic fields disappeared in adult animals. Only recently experiments on tissue recombination were interpreted as indicating that cell-to-cell, and stroma–parenchyma interactions operate in tissue maintenance during adulthood [9, 10].

Rubin proposed a modern version of the concept of morphogenetic fields as a substrate of carcinogenesis. In his own words, 'Cancer can be induced experimentally by disturbing the normal homeostatic relations among and between cells. . . . The perturbations could involve the integrated relations at the level of the organism, the tissue, or the cell. Once the ordering forces are disturbed, affected cells would express their innate capacity for heterogenization[1] which would in turn exacerbate the original perturbation in a positive feedback loop accounting for the progressive nature of neoplasia' [11]. Rubin's view is the closest to the tissue organization field theory of carcinogenesis. However, in Rubin's positivist criteria, only the observable phenomena should be introduced in a description of the phenomenon analyzed, with a minimal reference to premises that cannot be proven. In the tissue organization field theory, a fundamental premise is that the default state of cells is *proliferation*. When dissociated, cells are freed from organismal restrictions that compel them to express a phenotype appropriate to their position. Hence, they will exercise their built-in capacity to proliferate, and they may express new phenotypes, including their constitutive ability to migrate.

## On paradigm changes

When discussing the history of the 'cell theory', Bechtel called attention to the fact that even after Remak described the accrual of new cells in the frog egg by

[1] Heterogenization refers to the swift development of variation among cells that are dissociated and placed in tissue culture. This variation in the pattern of expression of markers, size, ability to propagate, etc.

direct cell division, the suggestion by Schwann–Schleiden that cells increased in number through a process of crystallization remained dominant for several years [12]. The prevalence of the somatic mutation theory of carcinogenesis, despite mounting evidence against it and the availability of alternative ways to interpret the accumulated data, is another clear example of the dictum 'the rejection of a theory does not rely on being contradicted by experimental data.'

As noted in Chapter 7, predictions about when the somatic mutation theory will finally collapse have not fared well despite the quality of the arguments presented and the stature of those who uttered those predictions. Not only scientific, but social factors as well, may have to be re-aligned to persuade scientists to concentrate their efforts on more promising leads. The emergence of a new *Zeitgeist* has at times been precipitated by a crisis developing in that particular area. We have made our case proposing the tissue organization field theory of carcinogenesis and neoplasia. This theory represents a synthesis between the negative hypothesis for the control of cell proliferation and the morphogenetic field theory of carcinogenesis.

## An integrative approach of control of cell proliferation and carcinogenesis

In *Table 1*, we compare the premises adopted to explore the subjects of control of cell proliferation and carcinogenesis. The somatic mutation theory is based on the proposition that *quiescence* is the default state of cells in metazoa, that carcinogenesis occurs at the cellular/subcellular level of complexity, and that neoplasias originate in a single cell. For supporters of this theory, the main features in carcinogenesis is an altered control of the cell number due to primary defects on the control of cell proliferation, of cell death, and of differentiation (the process through which stem cells generate their functional progeny). Hence, the somatic mutations that are supposed to cause carcinogenesis are those that mediate the effect of growth factors (their receptors, their signal transduction pathway), the cell cycle machinery (cyclins, cellular oncogenes, etc.), genes that mediate cell death, and those that form part of the maturation program of specific cell lineages. There are few genes left out of this list.

The tissue organization field theory is based, instead, on the proposition that *proliferation* is the default state of all cells, and that carcinogenesis takes place at the tissue organization level.

When exploring this theory the targets perceived as being relevant to carcinogenesis have to be reconsidered. First, it becomes irrelevant to invoke gene products that, stimulate the entry of cells into the cycle (growth factors,

**Table 1.** A comparison among competing hypotheses on the control of cell proliferation and theories on carcinogenesis

I. Positive hypothesis on the control of cell proliferation and the somatic mutation theory of carcinogenesis

Unstated premises

1. The default state for cells in metazoa is *quiescence*.
2. The control of the cell cycle is equivalent to the control of cell proliferation (note that this contradicts 1, see Chapter 1).
3. Carcinogenesis takes place at the cellular/subcellular hierarchical level of complexity.
4. Neoplasias are monoclonal.
5. Neoplasias arise when genes involved on the control of the cell number are mutated.

A. Control of cell proliferation
   a. Growth factors.
   b. Cell cycle effectors (cellular oncogenes, cyclins, certain suppressor genes such as Rb, p53, etc.).
   c. *Ad hoc* additions: inhibitory factors and suppressor genes.

B. Carcinogenesis. The somatic mutation theory
   a. Altered growth factor signaling pathway.
   b. Altered cell cycle effectors (cellular oncogenes, cyclins, etc.).
   c. *Ad hoc* additions: altered inhibitory factors and suppressor genes.
   d. Altered regulation of cell death.
   e. Altered differentiation pathway.

II. Negative hypothesis on the control of cell proliferation and the tissue organization field theory of carcinogenesis

Stated premises

1. The default state for all cells is *proliferation.*
2. Carcinogenesis takes place at the tissue hierarchical level of complexity.

A. Control of cell proliferation
   a. Colyones and colyogenes.[a]
   b. Cell-to-cell interactions.
   c. Tissue-to-tissue interactions.

B. Carcinogenesis. The tissue organization field theory
   a. Cell-to-cell interactions.
   b. Tissue-to-tissue interactions.

[a] These terms refer to physiological, cell-type-specific inhibitors of cell proliferation and the genes that code for them respectively [13].

oncogenes, etc.). Second, the characterization of signals that inhibit the proliferation of cells takes center stage in research programs on the control of cell proliferation. Finally, by positing that carcinogenesis takes place at the tissue level of organization, oncogenes and suppressor genes no longer are considered causes of carcinogenesis because they operate at the cellular/subcellular hierarchical level. At this level, so-called proto-oncogenes and some suppressor genes reacquire their roles as regular participants of the orderly sequence of events that take place during the cell cycle.

## The impact of our reassessment

The theories we proposed in this book point to three novel areas of inquiry. First, the development of a research program to define which are the nutritional requirements of metazoan cells and to determine which are the nutritive components of serum. Most importantly, since most cells cannot be propagated in a synthetic medium devoid of serum, this program will make it possible to experimentally ascertain which is the default state of metazoan cells. Second, while the development of the research program outlined above is under way, the adoption of the default state of *proliferation* automatically focuses the study of control of cell proliferation on the exclusive search for negative signals, that is inhibitors, spanning from cell to cell, extracellular matrix to cell, stroma to parenchyma, and endocrine mediators. Third, exploring the tissue organization field theory of carcinogenesis using the conceptual and methodological approaches taken by developmental biologists to study morphogenetic fields will finally fulfill the desideratum of introducing the study of carcinogenesis within the broad context of biology. This program will define where in the tissue organization field, and how, carcinogens act to induce neoplasias within the society of cells.

## And . . . finally

As you may recall, we stated in the Preface that while you were browsing through this book many of your cells were actively proliferating while the same amount, if not more, were not. Still others were dying a quiet, programmed death. We have reached another milestone in an intellectual journey that started over 150 years ago with Dumortier and the first version of the cell theory. Almost a century ago, researchers adopted the somatic mutation theory to explore carcinogenesis. For over four decades now, *quiescence* has been considered the default state of cells in metazoa. These options have been sufficiently explored and have explained neither the control of cell proliferation nor the development of neoplasms. In this book we presented an end-of-the-century assessment of the fields of control of cell proliferation and carcinogenesis. We offered alternative competing theories to explain these subjects, and, finally, outlined a strategy to explore the road for the scientific journey lying ahead. The options are these: to continue consuming the tranquilizer Peyton Rous alluded to four decades ago, to adopt the alternatives proposed in this book, or to go back to the metaphorical drawing board to find new ones. The choice is yours.

## References


1. **Carroll, S.B.** (1995) Homeotic genes and the evolution of arthropods and chordates. *Nature* **376**: 479–485.
2. **Mayr, E.** (1996) *This Is Biology.* Harvard University Press, Cambridge, MA.
3. **Buss, L.W.** (1987) *The Evolution of Individuality.* Princeton University Press, Princeton, NJ, pp. 69–117.
4. **Margulis, L.** (1975) Symbiotic theory of the origin of eukaryoric organelles. Criteria for proof. *Symp. Soc. Exp. Biol.* **29**: 21–38.
5. **Needham, J.** (1936) New advances in chemistry and biology of organized growth. *Proc. Roy. Soc. B.* **29**: 1577–1626.
6. **Waddington, C.H.** (1935) Cancer and the theory of organizers. *Nature* **135**: 606–608.
7. **Orr, J.W.** (1958) The mechanism of chemical carcinogensis. *Br. Med. Bull.* **14**: 99–101.
8. **Gilbert, S.F., Opitz, J.M. and Raff, R.A.** (1996) Resynthesizing evolutionary and developmental biology. *Dev. Biol.* **137**: 357–372.
9. **Tarin, D.** (1972) Tissue interactions and the maintenance of histological structure in adults. In: *Tissue Interactions in Carcinogenesis* (ed. D. Tarin). Academic Press, London, pp. 81–94.
10. **Cunha, G.R., Bigsby, R.M., Cooke, P.S. and Sugimura, Y.** (1985) Stromal–epithelial interactions in adult organs. *Cell Differentiation* **17**: 137–148.
11. **Rubin, H.** (1985) Cancer as a dynamic developmental disorder. *Cancer Res.* **45**: 2935–2942.
12. **Bechtel, W.** (1984) The evolution of our understanding of the cell: a study in the dynamics of scientific progress. *Stud. Hist. Phil. Sci.* **15**: 309–356.
13. **Sonnenschein, C. and Soto, A.M.** (1991) Cell proliferation in metazoans: negative control mechanisms. In: *Regulatory Mechanisms in Breast Cancer* (eds M. Lippman and R. Dickson). Kluwer, Boston, pp. 171–194.

# Index

Accessory sex organs, 44, 60, 62
Acetylcholine, 79
Achondroplasia, 49
Adenovirus E1A, 123
Anaplastic tumors, 92
Androgen, ix, 54, 60
    control of cell proliferation by, 62, 71–72
Aneuploidy, 126–127
Animal models, 7, 11, 21, 36, 38, 39, 40, 45, 49, 53, 63, 64, 65, 68–70, 72, 73, 84, 86, 93, 94, 103, 112, 118, 120, 121, 127, 128
Anthropocentricity, 7, 19, 136, 138
Antibiotics, 21
Antigens, 84–86, 87
Anti-oncogenes, 54, 105, 115, 123, 124, 125, 141, 142
Apoptosis *see also* Cell death 19, 32, 37, 61, 96, 118
    inhibition of, 47, 82, 83
Asbestos, 119
Ascitic fluid, 50
Assays, 47, 80, 81
    bacterial, 121
    bio, 43
    clonal, 83
    cytochemical, 37
Athymic nude mice, 66
Atrophy, 44, 60
Autocrine
    hypothesis, 63
    inhibitor, 51
    mechanisms, 47
Autoradiography, 38
Autotrophs, 15, 26
Auxotrophs, 15, 26
Axial plans, 55

B-cell receptors, 85
Bacteria
    proliferation rate, 15, 32
    reproduction, 15–17
BCSG *see* Brain cell surface glycopeptide
Beryllium, 119
Blastocysts, 125
Bone marrow, 81
Brain anomalies, 48
Brain cell surface glycopeptide, 50

Cadherins, 54
Cancer, 1, 99–109
    definitions, 100–101
    diagnostic methods, 92, 99
    etiopathogenesis, 138
    genesis, 51
    hereditary, 123
    human, 113, 120, 127
    tissue, 8, 103
Carcinogenesis, 54, 91–98, 100, 101, 103–104, 105, 106, 107, 112–130, 135, 139, 140–142, 141
    chemical, 120–122
    foreign-body, 118–119, 121, 128–129
    physical, 120
    viral, 122–123
Carcinogens, 92, 114, 117, 142
    chemical, 94, 104, 120–122, 127
    mutagenic effect, 106–107
    target of, 104, 109, 128, 129
Carcinoma, 48, 50, 104, 107, 119, 122
Cecidomyan gall midges, 18
Cell adhesion molecules, 54
Cell cycle, 17, 27, 33–37, 38, 115–117, 123, 140, 142
    control of, 8–9, 141
    definition, 8
    effectors, 122, 124, 141
    exponential phase, 33, 35, 37, 65
    G0 phase (quiescence), 8, 9, 16, 17, 27, 36, 37, 38, 49
    G1 phase, 8, 37, 38, 51, 53
    G2 phase, 8, 53
    genes, 72
    history of, 7–8
    kinetics, 8, 36, 47

M phase, 8, 32
markers, 17
maturation, 79, 83, 84, 140
molecular steps, 8, 9
phases, 7, 82, 116
regulation, 24, 37, 53, 70
S phase, 4, 7, 8, 32, 38, 39, 48, 51, 61
stationary phase, 15, 33, 34
velocity, 37
Cell death *see also* Apoptosis; Necrosis 15, 19, 32, 36, 37, 38, 44, 71, 96–97, 121, 140, 141
Cell default states, 16–18,
proliferation, 8, 9, 20, 26, 27, 36, 37, 39, 42, 44, 52, 53, 55, 64, 70, 72, 73, 74, 81, 84, 85, 86, 87, 100, 117, 129, 137–139, 140, 141, 142
proliferation in metaphyta, 26
proliferation in unicellular organisms, x, 9, 15, 16, 17, 26, 27
quiescence, 5, 8, 9, 14, 17, 20, 21, 22, 24, 25, 26, 27, 36, 39, 42, 44, 45, 46, 52, 55, 63, 64, 65, 69, 71, 72, 73, 83, 84, 85, 87, 104, 115, 117, 122, 138, 140, 141
Cell division, 2, 3, 4
Cell loss, rate of, 4
Cell migration, 47, 48, 102
Cell motility, 102, 118, 124, 139
Cell number *see also* Hyperplasia 4, 19, 37, 39
Cell nutrition, 26, 27, 36
hematopoietic stromal cells, 82
nutrient availability, 16, 17, 18, 19, 37, 46, 107, 137, 138, 142
nutritional requirements, 15, 27, 52, 55
role in control of cell proliferation, 20–26
role in regenerative process, 5
Cell proliferation *see also* Inhibition/inhibitory factors 1, 11, 19, 23, 32–33, 38, 39, 84–86, 87, 96–97, 115, 118, 121, 123, 126, 128, 129
arrest, 9, 54, 61, 62, 71
autonomy, 116, 117
history, 2–8, 41–59
hormonal effects on, 45
induction, 18, 69
limited, in culture, 25
marker, 63–65
measurement, 4, 7
methodological evaluation, 32–40
negative signal, 85
nutrition, 20–26
pattern, 20
quantitation, 22–23
regulation, 8, 23–24, 27, 49, 53, 55
stimulation of, 22, 24, 25, 27, 83
Cell proliferation control, 7, 8, 9, 11, 16, 19–20, 20, 36–37, 43, 44, 53, 54, 60–77, 84, 87, 96, 102, 115–116, 134, 135, 136, 138, 140, 141, 142
direct-positive hypothesis, 62–63, 64
indirect-negative hypothesis, 64, 65–68, 73
indirect-positive hypothesis, 64, 63–65, 68–69
negative-control hypothesis, 8, 9, 71, 53–54, 81–82
positive control hypothesis, 6, 8, 9, 45, 70
Cell proliferation rate, 4, 16, 33, 34, 71, 96
doubling time, 39
experiments, 33–36, 38
measurement, 7, 33–34, 35
regulation, 17
Cell starvation, 15, 16, 18, 36, 46, 50
Cell theory, 31
history of, 2–4, 11, 139–140
Cell-to-cell interaction *see* Interaction
Cell types
AKR-2B, 51
antigen-presenting, 84
B-lymphocytes, 84–86, 87
basal, 37, 47, 80
bone marrow, 81
brain, 50
BSC-1, 50, 51
cancer, 114
CCl-64, 51
*dco* mutant, 54
Ehrlich mammary carcinoma cells, 50
embryonal, 127
endometrial, 69–70, 72
epidermal, 47, 48, 81
epidermoid carcinoma A-431, 48
epithelial, 4, 48, 50, 60, 61, 69, 70, 71, 72, 73, 79, 117
EPO-target, 82
erythroid, 82, 83
erythropoietic, 38
estrogen target, 62–63, 66, 68, 69
*fat* mutant, 54
fetal, 26
fibroblast, 26, 50, 79
frank neoplastic, 107

germ, 138
granulocytes, 81, 83
heart, 83
hepatocyte, 6, 16, 43
keratinocytes, 47
kit ligand *see* stem
leucocytes, 26
liver, 16, 37, 41
lung, 51, 52, 83
lymphocytes, 47, 84–86, 87, 119
MCF7, 66–68, 72–73
mesodermal, 49
Moloney MuSV-transformed 3T3, 50
mono-potential, 79
multi-potential, 125
Mv1Lu, 51
myo-epithelial, 54
nerve ganglion, 47
neuron, 16, 24, 47, 48, 79
NIH 3T3, 50
non-neoplastic indictor, 50
p27 knockout, 116
platelets, 51
pluripotential, 79, 81
prostate, 62
Rb1, knockout, 116
red blood, 81
sex steroid target, 53, 73
spleen, 83
T-lymphocytes, 84–86, 87
teratocarcinoma, 96
totipotential, 79, 127
tumor, 50
uterine, 62, 70, 83
wild-type, 116
Cell volume, 4
Cell yield experiments, 35–36
Cells
age of donor, 25
developmental fate, 129
constitutive property (proliferation), 6, 15, 16
constitutive property (quiescence), 17
default state in metazoa, 9
informational content of position, 54, 81
self-renewal of, 79, 80
self-replication, 137
survival rates, 83
thawing, 37
transfected, 67
Cells in culture *see also* Cell types 8, 21–26, 27, 32, 38, 41, 45, 60, 63, 70, 72, 80, 87, 120, 127–128
apoptosis, inhibition of, 82
artifact, 25, 72, 73, 107
death, 37
history, 7
inhibition, 52
'malignancy', 107–108
metazoan, 33, 36
methodological problems, 106–108
methods to evaluate proliferation, 33
monolayer, 36
'progressive state selection', 107
proliferation rate experiments, 33–34, 39
propagation, 21
somatic mutation theory, 106
spreading factors, 52
staining, 33
suspension, 36
techniques, 45
'transformation', 106–107
transplantation, 21
validation of findings, 40
Cellulose membrane filters, 118
Central nervous system, 16
Chalones, 53–54
Chemotherapy, 100
Chloroplasts, 15
Chromatin, 7
Chromosomes, 7, 107, 113, 126, 127, 129
aberration, 106, 124
'chromosomal imbalance', 95
Chronic myelocytic leukemia, 126
Clonal expansion, 84, 85, 121, 123
Clonal growth, 46
Clonal selection theory, 84
CNS *see* Central nervous system
Cockayne syndrome, 120
Collagen, 6
Colony-stimulating factors (CSFs), 52–83
Colyogenes, 141
Colyones, 141
Contactinhibin, 50
Cranial sutures, 49
Crossing-over, 11, 18
Crystallization theory, 3
CSF *see* Colony-stimulating factors
Culture, 11, 65, 68, 72, 73
Culture medium, 14, 21, 24, 64
albumin, 23, 68, 70, 71
amino acids, 22, 23, 25
bioactive molecules, 24

defined, 20, 21, 23–24, 26–27, 35, 46
development, 14–15
estrogenless, 67
growth factor-free, 67
HGF-free, 44
serum-free, 24, 25, 26, 27, 46, 65, 67, 73
serum-supplemented, 21, 22, 23–24, 26, 35, 67
Culture models, 49
Cyclins, 70, 71, 140, 141
Cytokines, 83, 84, 85, 86, 87
Cytologism *see* Somatic mutation theory
Cytoplasmic processes, 24, 47
Cytoskeletons, 20

Decapentaplegic, 52
Definitions, 100
functional, 51
operational, 18, 46, 80, 81, 121, 134, 136
Demergence, 80, 107, 128
Descriptive research, 136
Descriptive terms, 5
Determination, 4
Developmental biology *see* Embryology
Developmental genetics, 109, 129
Developmental program, 19
Dextran-coated charcoal, 35
Dextran stripping, 66
Differentiated mass, 6
Differentiation, 4, 21, 79, 85, 86, 87, 126, 127, 140
cells, 4, 21, 80, 125
definitions, 78–79
factors, 47
hypotheses, 96–97
neoplasia, 92
pathways, 141
tissue, 78–87
Diploid organism, 16
DME, 24
DNA, 7, 39, 82, 94, 114, 118, 119, 120, 121, 122, 137
cDNA, 67
degradation, 19, 38
fragmented, 37
precursors, 32, 33, 37, 38
recombinant technology, 46, 67
sequences, 62
structure, 1, 23, 113
synthesis *see* Cell cycle, S phase
synthesis, induced, 4
transferral of fragments, 106
tritiated thymidine incorporation, 46
viruses, 123
Domain structure, 68
Down syndrome, 95
Drosophilia, 52, 54, 108, 116, 129
Dwarfism, 44, 49
Dysplasia, 54, 71, 104, 105, 107

EGF *see* epidermal growth factor
Electronic particle counter, 23, 33
Embryogenesis, 103, 113
Embryology, 3, 4
Embryonal cells, 127
Embryonal environment, 126
Emergent phenomena, 10, 11, 16, 94, 95, 97, 114, 117, 126, 135, 137–138
Endocrine
glands, ablation of, 44–45
hypothesis, 63, 65
mechanisms, 54, 55
mediators, 142
Endometrium, 19, 61, 69–70, 71, 72
Enucleated oocytes, 138
Epidemiological studies, 101
Epidermal growth factor, 47–49, 50, 52, 67, 69–70
Epidermis, 36, 37, 48, 53, 121
Epigenesis, 94, 107
Epigenetic mechanisms, 121
Epiphenomenon, 127
Epiphysial growth plate, 49
Epithelium, 4, 36, 48, 50, 54, 60, 61, 65, 69, 70, 71, 72, 73, 79, 80, 117, 121, 126
EPO *see* Erythropoietin
Erythrocytes, 38, 82, 136
Erythropoietin, 38, 52, 82–83, 87, 136
Estradiol, 61, 62, 63, 65, 66, 67, 68, 70, 71
Estrocolyone-I, 64, 67, 68
Estrogen, 60, 61–72, 73
control of cell proliferation by, 4, 62
estrogen-free, culture medium, 67, 73
receptors, 61–63, 67
receptor knockouts (ERKO), 69, 70
target cell lines, 62–63, 66, 68, 69
Euglena, 15
Eukaryotes, 14, 15–16, 26, 102, 137
Euploidy, 127
Experimental
approaches, 10, 94, 118
design, 2, 26, 36, 73, 135
models, 60, 66, 67, 72, 125–126, 128
Explants, epithelial, 65
Extra-cellular matrix, 6, 20, 125, 142

F10, 24
Fatty acids, 22
Fetal *see also* Embryonal
    cells, 26
    death, 82
    serum, 25
    tissue, 50
FGF *see* Fibroblast growth factor
Fibroblast growth factor, 49, 52
Fibrotic reaction, 119, 122
Flow cytometry, 32, 37
Frank neoplasia, 101, 104, 107

Gametes, 15, 16, 18
Gene, 128, 129
    cell cycle, 72
    cell death, 140
    concept, 11
    disabled genes, 96
    disruption, 48
    dorsal-ventral polarity, 52
    estradiol-induced, 70
    estrogen-receptor, 66
    estrogen-regulated, 69
    expression, hormonal effects on, 45, 48
    function, 46
    gas 1, 50
    gas 2, 50
    genetic defects, 95
    growth arrest-specific, 50
    inactivation experiments, 124
    induction, 19
    lethal giant larvae-2, 108, 129
    maturation programs, 140
    mRNAs, 62
    p53, 123, 124, 141
    P-cadherin, 54
    retinoblastoma, 123, 124, 141
Generative mass, 6
Genome, 4, 7, 114, 138
Germ plasm, 16
Glyco-proteins, 83
Gonadectomy, 60, 62
Growth
    arrest-specific gene products, 50
    catalysts, templates, 43
    clonal, 46
    control, 44
    definitions, 5, 6, 24
    history, 2–8
    hormones, 7
    regulation, 6
    retardation, 48
    stimulation, 6, 21, 117
    tissue, 2
    trophic signals, 45
Growth factors, 6, 7, 8, 9, 15, 17, 18, 21, 23–24, 25, 27, 34, 46–49, 67, 69, 70, 72, 73, 83, 85, 87, 106, 115, 117, 124, 136, 140, 141
    autocrine hypothesis, 63
    circulation, 49
    definition, 46, 47–8
    endocrine hypothesis, 63, 65
    epidermal growth factor
    erythropoietin, 38
    estrogen-induced, 65
    exponential phase, 65
    extra-cellular space, 49
    fibroblast, 50
    hepatocyte growth factor, 42, 43, 44
    insulin-like growth factor, 50
    insulin-like growth factor-I, 52, 69, 82
    interleukin-2, 86
    keratinocyte growth factor, 49
    mediators, 122
    mitogenic, 50
    nerve growth factor, 24, 47, 50
    nutrients, 15
    paracrine hypothesis, 63
    peptide, 51
    platelet-derived, 50
    receptors, 17, 46, 47, 48, 51, 115, 122, 124, 140
    recombinant species-specific, 46
    role within negative control hypothesis, 52–54
    transforming growth factor alpha, 44, 49
    transforming growth factor beta, 50–52, 116
    transforming growth factor beta-1, 44, 51, 52
    T-cell growth factor, 86

Hair follicles, 48
Hematopoietic stromal cells, 81, 82, 87
Hemoglobin, 124
Hemopoiesis, 79, 87
Hepatectomy, partial, 4, 42, 53
Hepatitis B, 122
Hepatomas, 127
Hereditary retinoblastoma, 123
Heterogenization, 139
HGF *see* Growth factors, hepatocyte growth factor

Hierarchies, 2, 6, 9–11, 43, 79–80, 86, 100, 116, 120, 121, 135
- cancer, 101–111
- carcinogenesis, 97–98, 124
- cell, 6, 19, 25, 36, 44, 46, 53, 97–98, 127, 128, 140, 141, 142
- definition of growth, 6
- extra-cellular matrix, 6, 20, 25, 49, 53, 125, 142
- integration of data, 2, 94
- intra-cellular, 6, 53
- neoplasia, 97–98
- organ, 44, 53, 128
- organism, 6, 32, 44, 45, 107
- subcellular, 19, 36, 127, 128, 140, 141, 142
- tissue, 6, 53, 79–80, 97–98, 126, 128, 140, 141, 142

Histogenesis, 39
History
- cell proliferation and growth, 2–8, 41–59
- cell theory, 11, 39–40
- cells in culture, 7
- study of neoplasia, 91–93

Hormones, 7, 22, 24, 53, 46
- effects on cell proliferation, 45
- effects on gene expression, 45
- purified, 45
- sex-hormone mediated control of cell proliferation, 60–77
- target organs, 96
- trophic role, 41, 44–45

Human model, 66
Hyperplasia, 5, 6, 54, 61, 71, 104, 105, 107
Hypophysectomy, 65
Hypophysis, 44, 45
Hypoxanthine-phosphoribosyl transferase, 95
Hypertrophy, 5, 6, 43, 61

Ideology, 134, 135–136
IL-2: Interleukin-2 (or T-cell growth factor) *see* Growth factors
Imaginal disks, 54
Immune reponse, 52, 84
Inhibition/inhibitory factors (of cell proliferation), 6, 8, 9, 34, 36, 37, 38, 39, 42, 43, 49–54, 55, 65, 81, 82, 83, 84, 87, 104, 105, 116, 117, 118, 123, 124, 136, 141, 142
- anti-templates, 43
- autocrine, 51
- blood-borne, 65
- mammary-derived growth inhibitor, 50
- neutralizers of, 84, 85–86, 105
- pathways, 70
- plasma, 42, 43, 66, 68
- plasma/serum borne, 64, 66–68, 72, 73
- reversal of, 68
- signals, 27, 45, 71, 124–142
- stromal cells, 82
- transforming growth factor beta, 116
- tissue-specific, 43

Initiation/promotion hypothesis, 120–121
Inoculum size, 33, 35
Insects, 18
Insulin, 46, 52
Integrins, 20
Interaction, 10, 20, 39
- cell-to-cell, 3, 19, 20, 51, 53, 54, 55, 82, 128, 138, 139, 141, 142
- disruption, 122, 124
- endocrine mediators, 20, 142
- extra-cellular matrix receptors, 20
- extra-cellular matrix to cells, 97, 142
- gap-junctions, 54
- intra-cellular, 43, 51
- junctional complexes, 20
- juxtacrine, 20
- paracrine, 20
- positional information, 54
- stroma-epithelial, 69–70
- stroma-parenchyma, 139, 142
- tissue-to-tissue, 141
- virus-host, 116

Interferons, 50
Intrauterine development of the liver, 44
Isoproterenol, 4

Karotype, 126
Keratin/keratinization, 24, 48, 79
Keratinocytes, 47
Kidney, 4, 5
Kinases, 116
Kit ligand *see* Stem cells
Knockouts, 124
- cytokine, 83, 86, 87
- EGF, 48
- estrogen-receptor 69, 70
- growth factors and receptors, 46, 69
- HGF, 44
- KGF, 49
- P-cadherin, 54

TGF-a, 49
TGF, b1, 52

Lag period, 33, 34
Lesch–Nyhan syndrome, 95
Ligand, 20
Light microscopy, 19, 32, 92, 101, 103
Limb defects, 49
Linkage effects, 11
Lipids, 23
Liver, 6, 16, 19, 37, 41, 81, 121, 127
intrauterine development, 44
regeneration, 4, 5, 41–44, 42, 53
Log phase, 33, 34
Lumen, 118
Luminal epithelium, 54
Lung maturation, 48
Luteinizing hormone, 45

Macrophages, 83
Malignancy, 107–108, 125
Mammary ducts, 54, 72
Mammary glands, 50, 61, 72, 121
Mammastin, 50
Mathematical models, 2, 6, 32
Maturing factor-beta, 50
Medical devices, 118
Meiosis, 16
Membranes
basement, 124
Menstruation, 19
menstrual cycle, 66, 67
Metabolic regulators, 46, 52
Metaphyta, 16–18, 21, 26
Metaplasia, 104, 105, 107
Metastasis, 92, 101, 102, 116, 117, 118
Metazoa, 16–18
Methodology, 11, 127–8
cells in culture, 41, 45, 106–108
evaluation of cell proliferation, 32, 40
tissue culture, 41, 45
Mitochondrial enzymes, 124
Mitosis, 7, 8, 32, 33, 47, 53, 81, 126
history, 3–4
mitotic cells, 4
mitotic index, 66
post-mitotic cells, 8
Monoclonality, 126, 127, 129, 141
Morphogenesis, 19, 37, 47, 52, 55, 69, 109, 126, 129, 139, 142
Morphogenetic field theory, 140
Morphometrics, 32

Mutation, 94, 106, 113, 114–115, 118, 120, 121, 124, 125, 127, 128–129
'gain of function', 49, 104, 106, 108, 115
germ-line, 95, 108
hypermutability, 120
'loss of function', 105, 108, 115, 116
reversion, 125
Myofibrils, 6
Myosin, 10

Necrosis, 19, 37
Neoplasia, 54, 71, 91–98, 100, 102, 103, 104, 108, 113, 114, 116, 117, 118, 119, 122, 123, 124, 126, 127, 128, 140, 142
cellular, 97–98
definitions, 91–92, 101
diagnosis, 103
emergence, 97, 126
frank neoplasia, 101, 104, 118
*'in statu nascendi'*, 126
mono-clonal, 141
revived theory, 103
study, historically, 91–93
tissue-based, 97–98
undifferentiated, 92, 127
Neoplasm, 96, 118
generation, 126
Neoplastic cells, 9, 22, 63, 92–95, 104, 113, 117, 119, 125–127, 128, 129
frank neoplastic cells, 107
non-neoplastic indictor cells, 50
Neoplastic development, 139
Neoplastic phenotype, 106, 118, 121
Neoplastic tissue, 50, 103
Neovascularization, 119
Neural development, 21
NGF *see* Growth factors, nerve growth factor
Normalization, 127
Nucleic acids, 7
Nucleosides, 22
Nucleotides, 37, 39
Nucleus, 3, 4, 33, 38, 80

Oligopeptides, 53
Oncogene hypothesis, 122
Oncogenes, xi, 49, 105, 106, 115, 117, 118, 122, 123, 124, 125, 136, 140, 141, 142
Oocyte, 18
cytoplasms, 80
Oogenesis, 138
Organ *see also* Hierarchies
cell numbers, 37

culture, 47, 67, 70, 128
growth, 2, 4–5, 6, 43, 63
mass reproduction, 43
organization, 20
regeneration, 2, 5, 41, 44
size, 19, 32
Organelles, 137
Organicism, 9, 114, 135, 136
Organism *see also* Hierarchies
diploid, 16
growth, 2, 4–6
regulation, 6
Organogenesis, 39, 129
Ovarian cycle, 61
Ovariectomy, 60, 61, 65, 70, 72, 73
Oxygen, 7, 19, 39

Papilloma viruses (E6, E7), 123
Paracrine hypothesis, 63
Paracrine mechanisms, 47, 54, 55
Parasitic disease, 119
Parenchyma, 6, 92, 42, 103, 104, 117, 127, 128, 129, 139, 142
Parenchymal cells, 92, 93, 97, 117
dissociation, 44
Parthogenesis, 18
Particulates, 18
Pathogens, 14, 15
Peptide chalone, 51
Peritoneum, 125, 126
Pfeiffer syndrome, 49
pH, 7
Phenol red, 66
Phenotype, 78–79, 80, 94, 95, 96, 106–107, 114, 117, 118, 129, 137, 139
abnormal, 48
discordant, 44
neoplastic, 106, 118, 121
specialized, 22
transformed, 50, 106
unequal, 4
Photosensitive tricothiodistrophy, 120
Physicalist-mechanicism, 113
*Pisum sativum*, 11
Pituitary glands, 61, 65
Pituitary tumors, 63
Placenta, 48
Plasma, 6, 23, 42, 43, 64, 66, 68
membrane, 20, 47, 49, 54
membrane receptors, 17
proteins, 23
Ploidy, 4, 126, 127
Polypeptides, 46, 47, 50
Polyploidy, 11
Primary culture, 67
Proestrus, 61
Progesterone, 61
Programmed cell death *see* Apoptosis
Progressive state selection, 107
Prokaryotes, 14–16, 23, 26, 102, 137
'Promoter', 121, 128
initiation/promotion hypothesis, 121
Prostate, 44, 45, 53, 62, 71
Proteases, 15
Protein synthesis, 48
Proto-oncogenes, 70, 71, 124, 142
Protoplasm, 3, 6, 43

Radiation, 120
amino acids, 23
ionizing, 106
precursors, 38
tracers, 7
Radiolabeled DNA precursors, 32
Radiolabeled estradiol, 61
Radiolabeled serum proteins, 23
Rb *see* Gene, retinoblastoma
Recombinant technology, 46, 68, 69, 73, 129, 139
Reductionism, 9, 10, 20, 94, 114, 119, 120, 135, 136
Regeneration, 2, 5, 6, 45
kidney, 5
liver, 4, 5, 41–44, 42, 53
Regression, 118
Reproduction, 15–16, 18
Respiration, 3, 124
Restrictions, 83
Reverse emergence, 10, 138
Reversion, 53, 107, 127
Ribosomes, 48
RNA viruses, 122

Salivary glands, 5
Saprophytes, 14
Sarcoma, 119
SCF *see* Stem cell factor
Schistosomiasis, 119
Self-replication, 1, 10, 138
Senescence, 25–26
Serratia marcensis, 15
Serum, 21, 36, 37, 46, 50, 64, 66, 72
concentration, 65
dialyzed, 23
estrogen-free, 73
fetal, 25

inhibitory role, 67, 72, 73
macromolecules borne in, 22, 24, 25
nutritive components, 142
nutritive role, 23, 27
proteins, 23, 36
starvation, 46, 50
stimulatory role, 25
supplementation, 22, 23–24, 26, 67
Serum-free medium, 24, 25, 26, 27, 65, 67, 73
Sex steroid, 7, 35, 53, 73
receptors, 60
Sialoglycopeptide inhibitor, 50
Signal transduction, 47, 115, 116, 122, 124, 140
Simian Virus, 40, 122, 123
Sinusoids, 16
Skin appendages, 48
Soma, 138
Somatic
cell genetics, 23, 24
cells, 4, 16, 25, 26, 114, 124, 138
components, 16
mutation, 96–97
Somatic mutation theory, 94, 95, 96–97, 104, 105, 106, 107–108, 113–117, 118, 119, 120, 122, 123, 124, 125, 126, 127, 128, 139, 140, 141,
methodological difficulties, 106–107
unanswered questions, 115
Spermatogenesis, 138
Spleen, 81
Spongiotrophoblast, 48
Spores, 15
Spreading factors, 52
Stem cells, 79–83, 87, 125, 140
apoptosis, 82
differentiated nature, 80
erythroid, 82
further classification of, 81
hemopoietic, 81–84
operational definition of, 80, 81
stem cell factor (also called kit ligand), 82
types, 79–80
Striated muscle contraction, 10
Stroma, 69–70, 73, 92, 93, 103, 117, 119, 121, 128, 129, 139, 142
Stromal cells *see also* Hematopoietic stromal cells 72, 82, 92, 93, 117
Submaxillary glands, 24, 47
Supernatant fluids, 50
Suppressor genes *see* anti-oncogenes
SV40 *see* Simian Virus, 40
Symbionts, 15
Syngeneic animals/hosts, 63, 93, 119, 128

Techniques/technical difficulties, 4, 5, 7, 21, 22, 45, 126
Template/anti-template theory, 43, 44
Teratocarcinoma, 96, 121, 125, 127
Terminal deoxynucleotide transferase, 37
Testosterone, 45
Thanatophoric dysplasia, 49
Thermophiles, 14
Thymidine, 4, 50
tritiated thymidine, 38–39, 46
Thymus, 81
Tissue *see also* Hierarchies; Neoplasia, definitions 107
aberrant patterns, 118, 128
differentiation, 78–87, 96
fetal, 50
growth, 2
growth factors, 63
induction, 55
maintenance, 103, 104, 129, 139
metazoan cell proliferation, 32–33
non-tumorigenic, 96
organization/disorganization, 6, 20, 92, 94, 104, 107, 124, 128, 129, 139
organization field, 119, 124, 126
proliferation rates, 4
quantitative studies, 32–33
recombination, 69, 129, 139
regeneration, 45
renewal, 36, 53, 87
tissue-specific inhibitors, 43
tissue-to-tissue interaction, 141
Tissue culture, 21–26, 41, 45, 128
Tissue organization field theory, 97, 105, 108, 117–118, 119, 122, 123, 127, 128, 129, 139, 140, 141, 142
forerunners of, 139
Tissue types
basal, 36, 37
bone marrow, 81
brain, 50
cancer, 8
connective, 117–118
embryonal, 21
epithelial, 36, 71, 107, 117–118
hematopoietic, 19
neoplastic, 50, 103
parenchymal, 92
stromal, 92

uterine, 69
Totipotential cells, 127
Toxic agents, 4
Transcription factors, 62, 122, 124
E2F1, 116
Transferrin, 25
Transformation, 106–107
Transmission genetics, xi, 3, 7
Transplantation, 21, 48, 63, 93, 96, 128, 129
Trophy, 41, 44–45, 50
Trypan blue, 37
Trypsin, 22, 33
Trypsinization, 7
Tumors, 50, 66, 91, 95, 96, 99, 100, 101, 106, 108, 112, 126
anaplastic, 92
definition, 101
generation, 112
rate of growth, 96
transplantation, 93
undifferentiated, 92, 127

Ultra-violent lights, 120
Undifferentiated cells, 4
Unicellular organisms, x, 16, 18, 32
Uterine epithelium, 4, 65, 69, 71
Uterus, 19, 44, 61, 65, 66, 67
interuterine, 44

Vagina, 61, 65, 66, 67
Vaginal epithelium, 69, 70, 72
Venules, 119
Vicia fava, 8
Vitalism, 10, 113
Vitamins, 22, 23, 25

Wild-types, 54, 70
*Drosophilia*, 108, 129

*Xeroderma pigmentosa*, 120

Yeast, 15, 17, 115

Zygote, 4, 9